Handbook of COTTON

G.K. Koutu
Jawaharlal Nehru Krishi Vishwavidyalaya
Jabalpur - 482 004

P.P. Shastry
College of Agriculture,
Rajmata Vijayaraje Scindia Krishi Vishwa Vidyalaya
Khandwa - 450 001

D.K. Mishra
Jawaharlal Nehru Krishi Vishwavidyalaya
Jabalpur - 482 004

K. C. Mandloi
College of Agriculture,
Rajmata Vijayaraje Scindia Krishi Vishwa Vidyalaya
Khandwa - 450 001

2012

Studium Press (India) Pvt. Ltd.

Handbook of COTTON

ISBN: 978-93-80012-39-1

Published by:

Studium Press (India) Pvt. Ltd.
4735/22, 2nd Floor, Prakash Deep Building
(Near Delhi Medical Association),
Ansari Road, Darya Ganj, New Delhi-110 002
Tel.: 91-11-43240200-15 (15 lines), Fax: 91-11-43240215;
jngovil@gmail.com; jngovil@hotmail.com

Printed at:

Salasar Imaging Systems, Lawrence Road, Delhi-35

FOREWORD

The cultivation of cotton in India has a very long history. Cotton and associated fibres provide employment for over 60 million persons. India is the only country where all the four species of cotton are under commercial cultivation. There has been a tremendous improvement both in the production and productivity of cotton since independence and more so during the last decade with the introduction of transgenic cotton. Further the efforts of Government of India through Technology Mission on Cotton has provided a definite impetus on all aspects of cotton cultivation, processing and marketing.

I congratulate the authors of the book for bringing out this very useful publication which provides an insight on the attempts made in the field of cotton improvement and all other related facets of cultivating cotton.

Gautam Kalloo
Vice Chancellor
JNKVV, Jabalpur

ABOUT THE EDITORS

Dr. Ganesh Kumar Koutu, Ph.D, is Principal Scientist, Plant Breeding and Genetics at JNKVV, Jabalpur, Madhya Pradesh. Dr. Koutu has obtained his B.Sc.(Ag.) in 1981, M. Sc.(Ag.) Plant Breeding and Genetics in 1983 and Ph.D. Plant Breeding and Genetics in 1988-89 all from JNKVV, Jabalpur, Madhya Pradesh. Dr. Koutu has joined his service as Assistant Cotton Breeder at Main Cotton Research Centre JNKVV, Khandwa. Dr. Koutu has served more than 20 years in cotton research and development. He has been engaged in cotton research with special emphasis on generating materials that has been widely used across the country to develop high yielding varieties of G. *hirsutum* and G. *arboreum*, hybrids with very high yields and excellent fibre properties. He has been associated in the development of first naturally coloured cotton variety JCC-1 in the *G. hirsutum* in India He has to his credit the development of arboreum varieties Jawahar Tapti, and JK 5, intra–hirsutum hybrid JKHy3 and hirsutum variety JK 4. He has more than 60 research papers in his credit in journals of national and international repute and has authored several bulletins. He is a member of the general body of MP Science and Technology, Bhopal and is also member of scientific advisory committee, Climate Change Board, Government. of MP, Bhopal. He is currently working as In-charge National Seed Project, JNKVV, Jabalpur.

Dr. P.P. Shastry born on 4th June, 1955 in Madhya Pradesh, Dr. Shastry did his schooling at Jabalpur and obtained an Honours degree in Microbiology from the University of Bombay in 1975. He did his Post Graduation in Plant Pathology from J.N. Agricultural University, Jabalpur in 1979. He started his career in the Department of Plant Pathology, JNKVV, Jabalpur in June 1980. He initially worked on betelvine focusing his attention on identification of strains and pathogenesis of Xanthomonas campestris pv. betlicola. He joined the College of Agriculture at Khandwa in 1989 and as a part of his doctoral work began working on the New Wilt of Cotton completing his Ph.D. programme in Plant Pathology in 1993 from JNKVV, Jabalpur. His pioneering work on understanding the intricacies of New Wilt has been widely appreciated. It was for this work that he was honoured with the prestigious HEXAMAR Award in 1993 jointly by the Plant Protection Association of India and the Hexamar Agricultural Research Foundation. His research paper on New Wilt was adjudged as the best research paper of the year published in Indian Journal of Plant Protection in 1995 and was honoured with the Smt. Sarada Memorial Award. He has extensively worked on another emerging problem of cotton–Myrothecium leaf blight. It was under his leadership that the Cotton Group at Khandwa was honoured by the District Administration in 2006 for their exceptional contribution in boosting the Cotton production and productivity. He was elected as a Fellow of the Indian Psychopathological Society, New Delhi in 1987 and the Fellow of the Plant Protection Association of India, Hyderabad in 2001. He was the Councilor of the Plant Protection Association of India (2000-03) and at present is the Councilor of the Cotton Research & Development Association, Hisar. Although Dr. Shastry is basically a Plant Pathologist, he has shown immense interest and has worked in close association of Plant breeders. Apart from assisting in developing several varieties of cotton, he has to his credit the development of an arboreum variety JK–5 with excellent fibre properties. Dr. Shastry was closely associated with the introduction of Bt Cotton in Madhya Pradesh and has been sensitizing all the stakeholders in Madhya Pradesh on various biosafety issues. He has recently (2009) visited the United States to study biosafety issues in genetically engineered crops. He has more than 50 research papers to his credit published in Journals of National and International repute and has authored several technical bulletins. He is a member of the Governing Body of the Madhya Pradesh Council of Science and Technology and is also a member of the State Biotechnology Coordination Committee of Madhya Pradesh. He is currently working as Dean, College of Agriculture, Khandwa under RVSKVV, Gwalior and is also the Head, Main Cotton Research Station, Khandwa.

Professor D.K. Mishra is Head of the Department of Plant Breeding and Genetics at J.N. Agricultural University, Jabalpur. He obtained his Bachelor degree in Agriculture in 1972 and Master degree with specialization in Genetics & Plant Breeding in 1974 from Banaras Hindu University, Varanasi. He joined J.N. Agricultural University, Jabalpur in 1974 and served in various capacities. He also performed his services as Associate Director Research Services, before leading the department. He has been elected as Fellow of the Indian Society of Genetics and Plant Breeding, IARI, New Delhi. In pursuance of academic excellence, he is credited for development of numerous crop varieties evolved for the nutritional and lively hood security of resources-poor farming community. He has been actively involved with the development of minor millet varieties benefiting the tribal community of the state. He has been a played major role in evolving three Kodo Millet, one Kutki millet and one wheat and one chickpea variety. There is more than 70 research communication of national repute to his credit. More than 15 students have been guided for post graduate thesis by him. He has been actively involved in the transfer of knowledge and technology among varied group of agricultural and certificate personnel since last 25 years. In his dynamic leadership the university has been credited for highest produce of breeder seed. The approaches evolved by him for maintenance of seed quality of all the Breeder seed at Jabalpur are of a prime significance.

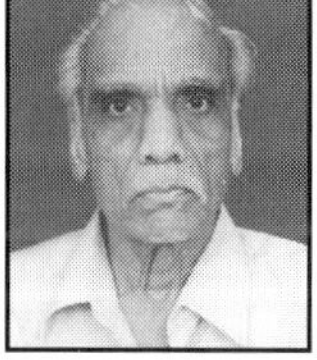

Kamal Chand Mandloi, retired Principal Scientist and Dean College of Agriculture, Khandwa, has distinguish academic record. Dr. Mandloi is Ph.D. in Genetics from IARI, New Delhi. He has guided several Ph. D. and M. Sc. students and published more than 60 papers and review articles in journals of national and international repute. Dr. Mandloi had widely travelled, renowned cotton breeder and associated in developing several varieties and hybrids of cotton (JKHy–2, JKHy–3 Jawahar Tapti and JK–4 etc.). Dr. Mandloi has been one of the stalwart cotton breeders of the country. The first naturally colored cotton variety in the *G. hirsutum* category in India is an unparallel contribution of Dr. Mandloi in the field of eco–friendly production of cotton in the country.

PREFACE

Cultivation of cotton and spinning cotton fabrics is known to India since ages. Asiatic (Diploid cotton) cottons were the first to be domesticated for cultivation. As such, cotton plays a major role in India's economy, both in terms of providing employment directly or indirectly to about 60 million people and in terms of production of wealth and earning foreign exchange for the country. Cotton improvement work was initiated in early 1920's and over the last 90 years tremendous improvement took place in almost all aspects of cotton cultivation in India. Although it was known that heterosis existed in cotton, the first commercial exploitation of heterosis was achieved in India of the World's first commercial hybrid–H4. Thereafter, several hybrids were developed and today area under commercial cotton hybrid cultivation in India occupies the largest in the world. Concerted efforts for systematic research and development (R&D) work in cotton started with the establishment with the establishment of the All India Coordinated Cotton Improvement Project by the Indian Council of Agricultural Research (ICAR) in 1966. Considerable expertise has been built up within the country in various disciplines relating to cotton production and the services of Indian Cotton Scientists and Technologists have been availed by various other countries.

As a result of systematic R&D work on cotton tremendous information on cotton research has accumulated. Several publications were brought to disseminate the accumulated knowledge. However, considerable progress in various fields have been made during the last four decades. It was in this context felt that there is a need to update the information and make it available to all concerned in cotton research.

This publication attempts to provide information on all aspects of cotton research with emphasis on cotton improvement, Plant Protection and quality parameters.

We take this opportunity to express our sincere gratitude to Dr. C.D. Mayee, Chairman, ASRB, Dr. Gopal Krishnan, Project Coordinator, AICCIP, Dr. K. Kranthi, Director, CICR for their guidance and views without which it would not have been possible to bring out this publication.

G.K. Koutu
P.P. Shastry
D.K. Mishra
K. C. Mandloi

CONTENTS

1

Introduction

Cotton has been from time immemorial and ever since domestication, an important commercial crop, whose outputs were ingeniously used to clothe the people and for feeding of livestock. Cotton, the versatile natural plant fibre with unique properties rendering it highly spinnable into yarns for textile use, is considered white gold because of the potential it offers to farmers and rural economy, global trade and national GDP. The United Nations Food and Agricultural Organization had devoted 2009 as the year of Natural Fibres all over the world, consequently its importance in global and national economies influencing farming and textile industries and other related sub-sectors were debated and discussed in various special fora in cotton growing countries including India and its blueprints were identified for positive growth in the 21st century.

From discovering the polyploidy mode through which the basic diploids and the allo–tetraploid cottons evolved in different continents along with a single gene mutation that caused lint hairs with special spinnable properties on seed epidermis. After that consequent domestication for economic exploitation promoted by gradual evolution of the mechanical ginning, spinning and weaving technologies in the industrial revolution eras in the 17th even and through 18th and 19th centuries. Gradually cotton cultivation and textiles have gone through brilliant revolutions in its cultivation, processing, trade and textile manufacture in the most successful era of the 20th century. With the efficient detoxification techniques in refining cotton seed oil for removing gossypol and other pigments, its value has further improved since the last few decades and added to that a larger output of cotton seed. By which it became possible to promote cotton through the commercial cultivation of *Bt*-transgenic varieties and hybrid cottons in the whole cotton world (Narayanan, 2010).

Cotton cultivation is a commercial farming venture in more than 85 countries with over 35–36 million hectares area and production touching some 26 plus million tonnes until the beginning years of the 21st century. In the last few years, the cotton area has shrunk to some 32–34 million hectares with a lower total production of some 23–24 million tonnes in the world (Narayanan, 2010). This is caused by various factors such as diversion of cotton area to bio-fuel crops in USA and other natural calamities like drought, floods, and competition by synthetic fibres like polyester etc., and production cost-price un-sustainability in various countries. The WTO has played an important role in trying to solve the problems like hidden subsidies and export subsidies by USA etc., causing setback to cotton production and trade in cotton and textiles by bringing the parties to the negotiating tables and find amicable solutions fair to all including cotton producers in rich, developing and poor countries.

World cotton production is projected by ICAC at 22.2 million tones in 2009–10, a 5% lower than in 2008/09 due to lower yield. This is the third consecutive season of decline in global cotton production, which resulted a decrease in price competitiveness of cotton *vs.* competing crops, and also weakening of cotton yields in the last two seasons. The production drop in this season is driven by a 16% fall in Chinese production, to 6.8 million tonnes. However, production is expected to increase by 5% to 5.2 million tonnes in India and by 8% to 2.1 million tones in Pakistan. U.S. production is forecasted slightly lower than last season, at 2.7 million tones.

India is better placed since the introduction of *Bt*-hybrid cottons from 2002–03 and the area under cotton has increased to around 9.6–10 plus million hectares, with an accelerated production of nearly 30.5–31.0 million bales (5.3 million tonnes) and average yield of 550–560 kg lint/ha along with 1100–1120 kg cottonseed/ha. India has increased its cotton consumption in textiles and other non-spinning uses to some 24.0 million bales and has achieved an export potential of 6–7 million bales of raw cotton. Exports of cotton yarn and textiles and modernization of the Indian textile industry has seen new milestones in the post-WTO liberalized trade era commencing from

Table 1.1: State wise cotton area (lakh ha) from 1999–00 to 2008–09

State	*99–00*	*00–01*	*01–02*	*02–03*	*03–04*	*04–05*	*05–06*	*06–07*	*07–08*	*08–09*
Punjab	4.75	4.74	6.00	4.49	4.52	5.09	5.57	6.07	6.41	5.37
Haryana	5.46	5.55	6.10	5.19	5.26	6.21	5.83	5.30	4.83	4.55
Rajasthan	5.83	5.10	3.47	3.86	3.44	4.38	4.54	3.50	3.68	2.23
Gujarat	15.39	16.15	16.90	16.34	16.50	19.06	20.77	23.90	25.16	24.17
Maharashtra	32.54	30.77	29.80	28.00	27.70	28.40	28.89	30.70	31.91	31.33
Madhya Pradesh	5.25	5.06	6.23	5.45	5.91	5.76	6.35	6.39	6.62	6.55
Andhra Pradesh	10.39	10.22	10.00	8.03	8.37	11.78	9.72	9.72	10.96	13.45
Karnataka	5.40	5.60	5.91	3.93	3.13	5.21	3.81	3.75	3.88	3.90
Tamil Nadu	1.85	1.93	2.00	0.85	1.03	1.29	1.52	1.22	1.30	1.20
Others	0.45	0.64	0.90	0.53	0.51	0.68	0.80	0.87	0.80	0.98
Total	87.30	85.80	87.00	76.7	76.00	87.9	88.20	91.40	95.55	93.73

Table 1.2: State wise cotton productions (lakh bales) from 1999–00 to 2008–09

State	*99–00*	*00–01*	*01–02*	*02–03*	*03–04*	*04–05*	*05–06*	*06–07*	*07–08*	*08–09*
Punjab	7.85	9.50	9.25	7.50	10.35	16.50	21.00	26.00	22.00	17.50
Haryana	10.7	10.00	5.50	8.75	11.50	15.50	14.00	16.00	16.00	14.00
Rajasthan	13.0	10.80	7.00	5.00	9.15	11.00	11.00	8.00	9.00	7.50
Gujarat	27.5	23.80	32.50	30.50	50.00	73.00	89.00	101.00	112.00	90.00
Maharashtra	38.0	18.30	34.30	26.00	31.00	52.00	36.00	52.00	62.00	62.00
Madhya Pradesh	15.50	19.30	20.00	18.00	19.65	16.00	18.00	18.00	21.00	18.00
Andhra Pradesh	22.50	25.30	26.80	19.80	27.4	32.50	30.00	35.00	46.00	53.00
Karnataka	7.00	7.75	7.00	5.00	4.20	8.00	6.50	6.00	8.00	9.00
Tamil Nadu	5.50	5.50	5.00	3.00	3.75	5.50	5.50	5.00	5.00	5.00
Others	1.50	1.00	0.75	1.00	1.00	1.00	1.00	1.00	2.00	2.00
Loose Supply	7.00	9.00	10.00	11.50	11.00	12.00	12.00	12.00	12.00	12.00
Total	156.00	140.0	158.00	136.00	179.0	243.00	244.00	280.00	315.00	290.00

January 2005 and the *Bt*–cotton production caused comfortable levels of cotton output. Cotton seed oil usage has almost doubled in the last decade (Narayanan, 2010). The state wise area production and growth trend in USA, China and India are presented in Tables 1.1 to 1.4.

Table 1.3: State wise cotton productivity (kg/ha) from 1999–00 to 2008–09

State	*99–00*	*00–01*	*01–02*	*02–03*	*03–04*
Punjab	281	341	262	284	381
Haryana	332	306	153	287	356
Rajasthan	379	358	343	220	379
Gujarat	304	250	328	317	475
Maharashtra	199	101	195	158	191
Madhya Pradesh	502	647	546	561	468
Andhra Pradesh	368	420	454	418	565
Karnataka	220	235	201	216	204
Tamil Nadu	505	484	425	600	613
All India	304	278	308	302	404
State	*04–05*	*05–06*	*06–07*	*07–08*	*08–09*
Punjab	551	610	728	583	554
Haryana	424	373	513	563	523
Rajasthan	427	397	389	416	572
Gujarat	651	794	718	757	633
Maharashtra	311	213	288	330	336
Madhya Pradesh	472	494	479	539	467
Andhra Pradesh	469	527	612	714	670
Karnataka	261	268	272	351	390
Tamil Nadu	725	668	697	654	708
All India	470	478	521	560	526

Table 1.4: Growth in Average yield (kg lint per hectare)

Year	*China*	*USA*	*India*	*World*
1980–81	550	453	169	411
1990–91	807	711	267	574
2000–01	1093	1008	278	612
2006–07	1245	894	520	733
2007–08	1225	976	555	794
2008–09	1265	985	526	785

Source: Anonymous, 2009–10

Cotton Growing Zones

The majority of the cotton in India is grown in ten states which are grouped into three different zones namely, Central zone (Maharashtra, Madhya Pradesh, Gujarat and Orissa), Southern zone (Andhra Pradesh, Karnataka and Tamil Nadu) and Northern zone (Punjab, Haryana, and Rajasthan). Approximately 65% of India's cotton is produced on dry land and 35% on irrigated lands (Table 1.2). Central and Southern cotton growing zones are predominantly rainfed. In 2009, of the total 9.6 million hectares, hybrids occupied 90% (8.6 million hectares) of the cotton area and only 10% (1.0 million hectares) were occupied by varieties. The percentage of hybrids has increased significantly with advent of *Bt* cotton, which have shown exceedingly over the non *Bt* hybrids.

Table 2: Cotton growing zones in India, 2008–2009

Zones	*North Zone*	*Central Zone*	*South Zone*
States	Punjab, Haryana, Rajasthan	Maharashtra, Madhya Pradesh, Gujarat, Orissa	Aadhra Pradesh, Karnataka, Tamil Nadu
Area	1.285 million hectares	6.121 million hectares	1.916 million hectares
Production	3.9 million bales	17.0 million bales	6.7 million bales
Productivity	516kg/ha	472kg/ha	594 kg/ha
Conditions	100% irrigated	Irrigated and rainfed	Irrigated and rainfed
Nature of genotype	Hybrids and varieties	Hybrid and varieties	Hybrid and varieties
Species	*G. hirsutum*, *G. arboreum*	*G. hirsutum, G. arboreum Intra hirsutum, G. herbaceum*	*G. hirsutum*, *G. arboreum* *G. herbaceum* *G. barbadense*, Interspecific tetraploids (HB)
Insect/Pest	*heliothis*, Whitefly, Jassids, Pink bollworm, Mealy bug	*heliothis*, Whitefly, Jassids, Aphids, Pink bollworm, Mealy bug	*heliothis*, Whitefly, Jassids, Aphids, Pink bollworm,
Diseases	Leaf curl virus, Wilt	Wilt	Wilt, Foliar disease
Sowing method	Drill Sown	Hand dibbling	Hand dibbling
Time of sowing	April–June	June–July	July–August

Source: Ministry of Agriculture, 2007 and Cotton Advisory Board, 2009

Constraints and Future Challenges in Indian Cotton

Considering the farming points of view, the Indian national average yields have significantly improved with the introduction of new gene technology of transgenic *Bt*-hybrid cottons in over 80 per cent of the cotton area. However, India is far behind from many advanced countries in terms of yield per unit area inspite of total production in the context of the largest area under cotton in the world. Yield instabilities and wide variations in contrast to input-output efforts, quality variations and deficiencies, problems to overcome natural vagaries and poor and unscientific management these are the factors that cause significant losses in realizable yields.

Certain cottons like those from *G. arboreum, G. herbaceum* and *G. barbadense* are required in significant proportions to the total cotton and at present, these cotton genotypes have become rare in the context of competitive emergence of *G. hirsutum*. *Bt*-hybrid cottons and their output are too low to maintain our time-honoured traditional textile trade and special end uses in textile manufacture.

Diverse cotton growing regions of India with varied production systems pose a challenge to the cotton improvement specialists. Drought, mismanagement of the crop by adopting unscientific practices, pests and diseases especially American bollworm (now manageable by *Bt*-transgenic varieties and hybrids), jassids, whiteflies and leaf curl virus disease, mealy bugs and micronutrient deficiencies etc., cause a heavy toll on the realizable yield levels. Mechanization of sowing with adjustment of plant spacing for optimum plant population in different soil types, plant type coupled with earliness and synchronized development of bolls and bursting, growth regulators and fitness of plant geometry for mechanical picking are some of the aspects posing a great challenge to cotton. Crop productivity gains for cotton in India have not reached a plateau unlike in certain other advanced countries. Annual rate of increase in yield from plant breeding-based advancements in cotton has been dramatic since the dawn of the 21st century. Considering the demonstrated potential globally of our current levels, the scope for doubling the current level of average productivity in the next two decades appears easy and convincing especially because of the emergence of newer genetic tools and techniques that are fast emerging. The stakes are high in the case of cotton seed oil also.

There for-which exclusive projects on basic germplasm identification and enhancement are essential (Narayanan, 2010).

STAPLEWISE PRODUCTION OF COTTON

Amongst the various fibre properties which influence the spinning performance or the quality, fibre length or staple length is most important. The basis adopted for classification of cotton into different staple length groups is not the same in all the countries due to differences in the level of estimation of staple length and types of cotton under cultivation. In our country, the classification in vogue is presented in Table 3.

Table 3: Staple length clasification in India

Category	*Staple length, Mean length range(in mm)*	*Approximate equivalent values (in inch)*
Short	19.0 and below	0.75 and below
Medium	20.0 to 21.5	0.80 to 0.85
Superior medium	22.0 to 24.0	0.85 to 0.95
Long	24.5 to 26.0	0.96 to 1.03
Extra long	27.0 and above	1.06 and above

Source: Anonymous, 1989

In 1947, with the partition of Indian subcontinent, area covering 40 per cent of the total cotton crop including a large proportion of long and medium staple cottons went over to Pakistan. Consequently, during 1947–48, cotton production in staple length above 22 mm in India was only 15 per cent which requires to import good quality cotton each year to meet the needs of the textile industry with considerable drain of foreign exchange. To overcome the shortage of good quality cotton and reduce the import of long staple cotton, research and development efforts were intensified for increasing the production of good quality cotton in the country. As a result, by the end of 1966 the country had able to produce more than 9 lakh bales of long and extra long staple cotton. After launching of All India Coordinated Cotton Improvement Project and after the development of varieties like MCU 5, Suvin and hybrids such as H 4, H 6, Varalaxmi and DCH 32, the long and extra-long staple cotton production was on the increase. At present we are producing as much cotton, which is more than our mills' requirement and thus have put us in a position to export this group of cotton. This could be possible only because of introduction of *Bt* cotton.

The Ginning Scenario in India

Available reports indicate that the first ginning and pressing factory in India was established in the present Vidarbha region of Maharashtra in 1868. Thereafter, their number seems to have grown rapidly over the years and at the time of partition in 1947, the Indian subcontinent had about 2000 units. After partition, India was left with only about 500 units having mostly gins. Subsequent increase in cotton cultivation area in different states and also development of high yielding varieties and hybrids which boosted up cotton cultivation and also saw a corresponding increase in ginning capacity in all the cotton growing states in the country.

The total number of saw, double and single roller gins installed in the factories is about 500, 19,000 and 30,000, respectively (Table 4). As stated earlier, most of the saw gins are used in the northern states of Punjab, Haryana and Rajasthan, where short and medium staple cottons are mainly produced. Single and double roller gins are exclusively used in the central and southern states. The total installed capacity is quite adequate to process all the cotton produced in the country.

Indian Textile Industry

The textile industry was the pioneer in industrialization. It occupies a pivotal position as the clothier of the nation and an important foreign exchange earner. It employs about 11 lakh workers in 2568 textile mills, 60 lakh workers in decentralized powerlooms, about 44 lakh workers in handlooms, 60,000 workers in manmade cloths, and 50,000 workers are employed in cotton farming and processing, manufacturing of dyes and chemicals and production of textile stores and accessories (Radhakrishnan, 1999). In addition textile related employment is provided in tertiary sectors like transportation, banking, insurance and marketing of textiles and allied products both at home and abroad. The total employment, direct and indirect offered by the textile sector would be in the vicinity of 60 million people.

Table 4: State–wise details of ginning and pressing factories in India

State	*Number of ginning and pressing factories*					
	Ginn-ing	*Press-ing*	*Compo-site*	*Total*	*Type of gin*	*Number of gins*
Punjab	137	12	94	243	SG	240
					DR	834
					SR	3008
Haryana	37	3	52	91	SG	115
					DR	114
					SR	2552
Rajasthan	220	8	59	287	SG	100
					DR	506
					SR	1701
Gujarat	1002	30	159	1191	SG	10
					DR	506
					SR	15810
Madhya Pradesh	246	14	50	310	DR	1387
					SR	2412
Maharashtra	387	16	135	538	SG	12
					DR	4477
					SR	3010
Andhra Pradesh	446	12	67	525	SG	3
					DR	5785
Karnataka	368	32	28	428	DR	2653
Tamil Nadu	394	9	16	419	DR	2034
					SR	1000
Total	**3237**	**135**	**660**	4032		

Source: Anonymous, 1999

The textile industry enjoys a weight age of 18% in the index of industrial production in the country. The industry's share in the Gross National Product is over 6%. The industry contributes over Rs. 53 billion to the central exchequer by way of excise revenue and accounts for 10% of the total excise earnings. In addition, there is substantial contribution from the textile sector towards state and local taxes and levies. Annual export of textile products has crossed Rs. 44000 crores and forms 30% of the aggregate foreign exchange earnings of the country (Radhakrishnan, 1999). Thus the sweep of the textile industry in the national economy is deep and widespread. The data on fibre consumption by textile industry and growth trend of Indian textile machinery are presented in Tables 5 and 6.

Table 5: Fibre consumption by textile industry

Fibres	***Textile industry's consumption***
Cotton	2,497 (78%)
Viscose staple fibre	173 (5.4%)
Polyester fibre	444 (13.9%)
Acrylic fibre	80 (2.5%)
Other misc. fibre	6 (0.2%)
Total	**3,200 (100%)**

Source: Anonymous, 1999

Table 6: Growth trend of Indian textile machinery industry

Machinery category	***1995–96***	***1996–97***	***1997–98***	***1998–99***
Spinning	885.09	652.88	672.17	514.25
Synthetic fibre	119.67	126.86	272.46	90.64
Weaving	118.77	108/13	98.34	72.63
Processing	90.78	90.53	98.02	102.07
Misc.spg., wvg. and Processing	1.97	1.75	3.86	4.42
Textile testing/Measuring instruments	29.89	34.22	29.56	25.22
Hosiery machinery and Needles	37.54	35.85	43.97	58.72
Spares and Accessories	216.60	240.72	282.33	269.55
Grand Total	1500.31	1290.94	1500.71	1137.50

Source: Anonymous, 1999

Textile Exports

Textile exports account for a third of the total foreign exchange earnings of India. The total exports of the textiles and clothing in 1998–99 has crossed Rs. 44,290 crores of US $ 10.50 billion. Even so, India's share in the total world trade in textiles continues to hover around 3%. However, in the case of cotton yarn a major breakthrough has been achieved and the country exports about 20% of the cotton yarn production.

The garment and hosiery sectors have also been exporting stridently in the last few years. The current value of annual exports (1998–99) of clothing is Rs. 22,375/– crores.

Coloured Cotton

The commercial cultivation of naturally coloured cottons, especially as "Eco-Friendly Cotton" has assumed special importance in some western countries to meet the challenges and health hazards increasing due to the effluent water of textile industries causing severe pollution of drinking water resources. It is needless to mention that the chemical dyeing process employed for the dyeing of fabric is discharging millions of gallons of highly toxic effluent water everyday containing very high concentration of heavy metal like Hg, Pb, As, Fe, Co etc., which is damaging the environment day by day. It is not merely a question of fashion of craze for fabric of naturally coloured cotton or of earning foreign exchange by creating attraction for coloured cotton, which is encouraging the research and development of coloured cotton around the world, but it is rather really a matter of concern for the biologist and agricultural scientist to realize the importance of naturally coloured cotton to protect the global environment and minimize/eliminate dangers of water and air pollution. However, the white cottons need due care and protection legislatively against any type of mechanical or genetical contamination. The research and development efforts on existing colour cottons are to be considered very important, of course not at the cost of white cottons, when the mankind has to choose the alternative whether to live and let live or continue the existing pollution creating modernized industrial progress without caring for the future generations.

It is in this context that exploitation of naturally coloured cotton assumes special significance as a sign of awareness for at least protecting drinking water for the mankind in future as well as for the life of animals, birds and aquatic creatures in the larger interest.

Germplasm resources of coloured cottons are available in National Gene Banks of USA, India, China, Uzbekistan, etc. In India, about 40–60 coloured cotton genotypes (genetic stocks) are reported to be available besides those of wild species and races (Narayanan and Sundaram, 1996). These represent both indigenous and exotic accessions from USA, Central Asia, Isreal, Peru, Mexico, Egypt, etc.

The coloured types in Indian gene resource include Algerian Brown, Arvin Brown, Texas Rust, LL.55–68–2, LC.1–1, Hirsutum Tashkant, Parbhani American, LC1–3, Brimer Browen, Lousiana

Brown, Fuzzy Seed Brown, SA.65, LC.1, Red 5–7, Brown Hirsutum, Cotton 162, Cotonnark Dark Brownand Light Brown, Kampala Red, Nankin Brown, Nankin Khaki, Extreem Okra, Arkansas Green, Arkansas Brown, Texas Brown, Intense Red Green Lint, KC. 94–2, KC–94–2–28 etc in *G. hirsutum*, Egyptian Brown, Green Lint, Pima Brown, etc in *G. barbadense* and Cocanadas 1 and 2, Khaki Arboreum, Red northerns, RH25, etc in *G. arboreum*.

Ms. Sally Fox (1987) is the first commercial cotton breeder who have developed American colour linted genotypes/cultures which are claimed to be comparable in all characters to white cotton. She has developed varieties with brown, green and red coloured lints and also multi-coloured lint such as green and white, off white and tan, and red and green. The commercial lines developed and patented for sale by Natural Cotton Colours Inc., are also becoming popular. However, no viable market has been established in USA for organic coloured cotton.

In India the work on development of naturally coloured cotton genotypes in *G. hirsutum* having high yield potential, resistance to pests and diseases with superior fibre and colour properties for spinning and fabric manufacturing purpose undertaken at JNKVV, Khandwa, has resulted in development of genotype KC–94–2 (JCC.1). This genotype was evaluated in large scale for commercial production in M.P. The yarn of 30s count produced in small scale trials was found suitable for weaving into a shirting fabric with 66 ends/inch and 46 picks/inch.

Organic Cotton

Traditionally, Indian farmers have been following organic cultivation till the middle of the last century, as they had no other choice. The advent of high yielding varieties and hybrids of cotton during the last three decades has resulted in large scale dependence on chemical fertilizers and especially toxic pesticides. Cotton, cultivated on 5% of India's cultivable land consumes 54% of the total pesticides used in Indian agriculture (Mehta, 1996). This has not only increased cost of cultivation but also caused environmental pollution. Some of the chemicals marketed, particularly, for plant protection are highly dangerous to human health and also leave poisonous residues in soil after application lasting over a few years.

Introduction

Global awareness of health and environmental issues is spreading fast in the recent years, especially in the developed countries. Sustainability in production has become the prime concern in agriculture development. Organic method of farming is the best option to ensure air, water and soil around us unpolluted leaving the environment safe for the present and future generations

A new trend is being developed in India now to produce some of the crops organically not only on account of the love for protection of nature but also because of the need for having safe products for human consumption.

Cotton enjoys an extremely positive image due to its naturalness and gentleness to human skin. Life cycle of cotton products from cultivation through production to disposal is embedded with harmful chemicals, which seems to impair human health and develop allergic illness. Realizing that millions of tones of chemicals are used to produce and process cotton and the likely effects these chemicals will have on human health and environment in the long run, it becomes imperative that organic cotton production is enhanced.

The 68th Plenary Meeting of the International Cotton Advisory Committee (ICAC) suggested that organic cotton has the potential to provide new idea that can influence and support wider sustainability in the sector. The organic cotton sector continues to grow, albeit more slowly in the global economic slowdown, reaching over 180,000 metric tonnes of lint in 2008/09 (Narayanan, 2010). Organic cotton is one of the options to meet the challenges of land use, food security and water security. Small growers exposed in the variable rainfall and problems from debt produce most organic cotton in India. Many farmers view organic cotton as a risk reduction tool, but the use of fewer purchased inputs involves a trade-off with productivity. Often the smallest and most resource poor farmers come to organic, and the fact organic works better in marginal or tribal areas. Certification of organic cotton is expensive and in some states of India the government is supporting certification costs. Organic cotton production, processing, spinning and marketing at retail level involves a long and complex chain requiring traceability and communication links between producers and consumers. The future of organic cotton production may involve contract farming with direct link to retailers.

Adoption of Bt *Cotton Hybrids in India*

Bt cotton, which confers resistance to important insect pests of cotton, was first adopted in India as hybrids in 2002, There were 54,000 farmers which grew approximately 50,000 hectares of officially approved *Bt* cotton hybrids in 2003. The *Bt* cotton area increased again many fold in 2004 to reach half a million hectares. In 2005, the area of *Bt* cotton in India continued to high scale reaching 1.3 million hectares, which is an increase of 160% over 2004. In 2006, the adoption of *Bt* cotton admitted a record increases which continued with almost a tripling of the area of *Bt* cotton to 3.8 million hectares. Remarkably in 2006, India's *Bt* cotton area increases from 3.8 to 6.2 million hectares, to become the largest area of *Bt* cotton in any country in the world. During 2007–08 cotton was grown on 9.44 mha, of which more than 80% was sown to *Bt* cotton. Cotton production in this year was 31.5 million bales (mha) (Anonymous, 2009–10).

During 2008–09 India is heading towards unprecedented record crop for sixth consecutive year after introduction of *Bt* cotton. There has been reduction in cotton acreage by 2 per cent in 2008–09 and stands at 9.3 million hectares compared to last year (The Cotton Corporation of India, 2008). Indian government has approved 263 *Bt* cotton hybrids till date. The area under cotton during 2008 has come down because of initial deficient rainfall in two major cotton growing states like Maharashtra and Gujarat.

References

Anonymous (1989). Technology for increasing cotton production in India. AICCIP, CICR, Regional Station, Coimbatore. p. 15.

Anonymous (1999). Hand book of cotton in India. ISCI Pub. pp. 1–500.

Anonymous (2009). National Cotton Scenario. Ministry of Textile.

http:/www. Cotcorp.org. in/national_cotton.org

Anonymous (2009–10). Annual Progress Report of AICCIP, Main Cotton Research Centre, Khandwa. MP. pp. 1–110.

Basu, A.K. (1996). Need for suitable policy and legislative measures for commercial cultivation of naturally coloured cotton. *J. ISCI*. pp. 202–206.

Fox, Sally (1987). Naturally coloured cotton- *Spin-off*, 46: 29–31: In search of coloured cotton- *Spin-off.*, 46: 48–50.

Iyer, K.R.K., Iyer, P.B., Sundaramurthy, V.T., Mandloi, K.C. and Meshram, L.D. (1996). Characteristics of coloured cottons and aspects requiring improvement. *J. Indian Soc. Cotton Improv.*, 21: 207–213.

Joshi, M.N. (1196). Domestic and export trade prospects of organic cotton. *J. ISCI* (Special issue on Eco-Friendly Cotton). 21(2): 234–237.

Mandloi, K.C., Khadi, B.M., Shirsat, Sidharth and Kulkarni, V.N. (1999). Naturally Coloured Cotton 9.397. Handbook of Cotton in India. Indian Society for Cotton Improvement, Mumbai. p. 548.

Mandloi, K.C., Iyer, K.R.K., Julka, G.K. and Julka, R. (1996). Development of coloured American cottons. *J. ISCI* 21(2): 180–188.

Mehta, K. (1996). Production and Certification of Organic Cotton. *J. Indian Soc. Cotton Improv.*, 21(2): 141–146.

Narayan, S.S., and Sundaram, V. (1996). Basic requirments and breeding procedures for developing naturally coloured cotton in India. *J. Indian Soc. Cotton Improv.*, 21(2): 159–169.

Narayanan, S.S. (2010). Reorientation of cotton improvement models and strategies in practical perspectives for 21st century. *Cotton Research Journal, ISCI* 1(1): 1–17.

Radhakrishnan, C.V. (1999). Indian textile Industry. Handbook of cotton in India. ISCI Pub. pp. 463–475.

2

Hirsutum and *Arboretum* Cotton Improvement

Credit for initiating Indian cotton improvement programme goes to East India Company who in late eighteenth century introduced American cottons to improve quality and quantity of our cottons. Systematic cotton improvement programme was however launched in the beginning of twentieth century with several states embarking on cotton research and development. Then in 1921 Indian Cotton Committee was established by Govt. of India to fund and coordinate cotton research of the country. ICCC sponsored several hundred research schemes pertaining to breeding of new varieties, agronomy, disease and pest management and marketing which helped in improvement of cotton. It was ICCC which established Cotton Technological Research Laboratory (renamed as Central Institute for Research on Cotton Technology) to evaluate fibre quality of new strains and breeding material as well as to conduct basic research in cotton fibre technology. ICCC was however abolished in 1966 when Indian Council of Agriculture Research took over the responsibility of research in all crops including cotton. It was in 1967 that All India Coordinated Cotton Improvement Project was launched by ICAR to improve both quality and quantity of cotton looking into the needs of domestic textile industry and export. Concerted efforts by various AICCIP centres such as Ludhiana, Faridkot, Sriganganagar, Hisar, Surat, Khandwa, Indore, Akola, Nanded, Parbani, Rahuri, Guntur, Dharwad, Raichur, Coimbatore, Srivilliputhur, Kovilpatti etc., for cotton improvement and strides that country made in cotton production for the last three decades or more are well recorded in the history.

The spectacular increase in production and productivity is mainly due to the intensive research programme, that are under way at various AICCIP centers. More than 200 varieties belonging to *G. hirsutum*, *G arboretum* and *G. herbaceum* have been released for commercial cultivation during the past three decades. The important milestone achievements are:

- The Jassid resistant medium staple variety Khandwa 2 from Madhya Pradesh and SRT 1 from Gujarat which later paved the way for the development of several popular hybrids.
- Development of Leaf Curl Virus resistant varieties RS 810, F 1861, LH 1556 and H 1098 (Anon, 2004) have helped the north zone to manage the disease incidence effectively.
- Development of long linted, high yielding *G arboretum* varieties such as Parbani Turab (PA 255), PA 405 from Parbhani, DLSA 17 from Dharwad and Jawahar Tapti and JK 5 from Khandwa have helped the farmers of south and central zone to grow this neglected species.
- Development of extra long staple variety MCU 5 and MCU 5VT and Surabhi is another landmark in the history of AICCIP.
- Development of F 414, H 777 with medium duration led to adoption of cotton-wheat rotation in north zone.
- Widely adaptable medium staple variety LRA 5166 which occupied large area in the country

Cotton is being cultivated in India in three distinct agro–climatic zones *viz.* northern cotton zone (N.Z.), central cotton zone (C.Z.) and southern cotton zone (S.Z.). The northern zone, comprising the states of Punjab, Haryana, northern Rajasthan and western U.P., occupies an eminent place on the cotton map of the country.

The present prominent position of cotton in the north zone has been earned through the development of improved varieties and better crop production and protection technology.

Achievement in Cotton Improvment Research

North Zone

Hirsutum

The American cotton (G. *hirsutum* L.) which made its humble beginning from isolated plants surviving as mixtures with arboreum cottons in the beginning of the twentieth century has now well established. The yield potential and adaptability of hirsutum has been considerably improved as a result of which about 90% of the area under cotton in the zone is now covered from about 20q/ha to over 45q/ha. The LSS variety released in 1933 which covered substantial area before partition was the only variety available for cultivation in this zone in 1947. It covered only 18.000 hectares and was cultivated as a single crop. Subsequent varieties like 216F, 320F, H 14, LL54, J 34 and J 205 had slightly better yield potential. The evolution of F 414 and H 777 in 1970's with medium duration (180–190 days) and high yield potential led to the adoption of cotton-wheat rotation on a massive scale. Since then, the duration of 180–190 days has become an essential feature of all the newly developed varieties.

A further gradual improvement in yield potential in this maturity group is witnessed by the evolution of several other varieties like LH 372, Ganganagar Ageti, F 286, H 655_c, F 505, BC 761, P 31, LH 886, HS 45 and LH 1134 (Table 1). The highest step in yield potential was achieved through the development of LH 900 which shows simultaneous increase in yield potential (up to 45 q/ha) as well as reduction in duration (150–155 days) over other varieties. An increase of yield from 20 q to 45 q and reduction for maturity period from 250–270 to 155 days is indeed a hallmark in the field of cotton breeding.

Arboreum

The yield potential of arboretum cottons has also been improved from 18 q to 38 q/ha. The first popular variety was 231 R which was replaced by G 27 in 1970's. since then very high yielding varieties like LD 133, RG 8, DS 1, DS 5 and LD 327 have been developed. Out of which LD 327 has displayed the highest yield potential in the north zone.

Table 1: Important varieties released in the north zone

Species/ Variety	*Year of release*	*Potential yield (kg/ha)*	*Ginning outturn (%)*	*2.5% span length (mm)*	*Spinning potential (HSC)*	*State to which recommended*
			Hirsutum			
J 205	1973	2000	34.3	25.1	40's	Punjab
B.N.	1976	2000	35.5	23.2	30's	Rajasthan
F 414	1977	2000	34.2	23.6	30's	Punjab
SH 131	1977	2000	34.0	21.5	30's	Uttar Pradesh
H 777	1978	2800	34.6	23.6	30's	Haryana
H 655C	1978	2500	33.3	27.4	40's	Haryana
G.A.	1978	2500	36.0	25.4	30's	Rajasthan
LH 372	1980	2300	33.3	25.8	30's	Punjab
F 286	1983	2800	33.8	23.8	30's	Punjab
LH 900	1985	4500	33.9	24.0	30's	Punjab
BC 761	1985	2500	35.0	24.5	30's	N.Z.
F 505	1986	3000	34.6	25.0	30's	Punjab
Vikas	1987	2000	33.0	23.5	30's	Uttar Pradesh
LH 886	1988	3800	35.0	25.4	35's	Punjab
HS 45	1988	3000	34.0	24.1	30's	Haryana
P 31	1988	2500	35.0	27.0	40's	N.Z.
LH 1134	1990	4200	35.5	27.6	40's	Punjab
RST 9	1991	2600	36.0	23.6	30's	Rajasthan
RS 810	2001	2500	35.0	27.0	35's	Rajasthan
Vagad Kalyan	2001	2500	35.0	27.0	35's	Rajasthan
H 1117	2002	2000	35.0	25.0	35's	Haryana
			Arboretum			
G 27	1969	2000	37.0	15.5*	–	Punjab, Haryana
Lohit	1969	1500	39.0	18.0	–	Uttar Pradesh
LD 133	1978	2000	38.9	17.0*	–	Punjab
LD 230	1981	2800	37.8	18.0*	–	Punjab
DS 1	1985	2200	39.0	16.5	–	Haryana
RG 8	1985	2500	40.1	16.5	–	Rajasthan
LD 327	1987	3500	41.9	17.5	–	Punjab
DS 5	1987	2500	40.4	16.0	–	Haryana

Source: Anon (2004), Anon (1995)

Most of the earliest forms of hirsutum cotton including LSS, 216 F, 320 F and H 14 were susceptible to jassids and bacterial blight. The lack of resistance to jassids and the bollworms wiped out cotton from central districts of Punjab in sixties. LL 54 variety failed only due to its susceptibility to bacterial blight. At least 2–3 sprays of systemic pesticides were required to protect the crop from the attack of jassids. But all the varieties now being cultivated in the north zone are fairly resistant to jassids as a result of which the use of systemic insecticides has been reduced from 3 sprays to only one and that also if required. Similarly, the incidence of bacterial blight has been considerably reduced through the development of resistant varieties. All the varieties except Ganganagar Ageti released after LL 54 have demonstrated their potential to resist the attack of bacterial blight. The incorporation of high degree of resistance to jassids and bacterial blight has reduced the risk of crop losses considerably.

Breeding Achievements in Central Zone

The central zone comprises the states of Gujarat, Maharashtra and Madhya Pradesh besides south Rajasthan, northern part of Andhra Pradesh. Bihar, Assam and Orissa have also been included in this zone for convenience. Largest area (nearly 50%) of the national cotton area lies in the zone. The zone is predominantly rainfed, characterised by early excessive rainfall, undulating topography of soils, marginal and shallow soils, poor farmers and low input application.

There have been excellent efforts at AICCIP centres to improve the fibre characteristics of *G. arboretum* and *G. herbaceum* through introgression of genes that increase the fibre length and strength. *G. arboretum* varietiers such as Jawahar Tapti and JK-5 from JNKVV Khandwa, Parbani Turab (PA 255) and PA 405 from MAU, Parbani, DLSA 17 from UAS Dharwad, etc. have been develop (Table 2). This silent revolution of diploid improvement is noteworthy during the last decade of twentieth centuary in AICCIP.

BREEDING ACHIEVEMENTS IN SOUTH ZONE

The south zone of AICCIP comprises Tamil Nadu, Karnataka and Andhra Pradesh. The list of varieties released in all the four cultivated

Table 2: Improved varieties released under AICCIP

Variety	*Year of release*	*Seed cotton yield potential (q/ha)*	*GOT (%)*	*M.F.L (mm)*	*Spinn-ing (HSC)*	*Tract/State to which recomm-ended*
		***Hirsutum* (American cotton) Varieties**				
Khandwa 1 (A 56-347)	1969	6–8	36.0	24.0	38s	Rainfed Nimar (MP)
Khandwa 2	1971	7–9	36.5	24.0	36s	Nimar (MP)
DHY 286	1975	14(7R)	36.0	28.0	50s	Vidharbha (MS)
G.Cot 10 (SRT1)	1975	14(8R)	36.5	23.5	40s	Mid and South Gujarat (also later M.S.)
G. Cot 100 (Vishnu)	1975	25–27	32.2	29.0	50s	Gujarat
Vikram (76 IH20)	1981	9	36.0	22.5	40s	Malwa and Nimar (MP)
G. Cot 12 (Kapatia)	1981	18	36.0	24.0	24s	Mathio tract (GS)
Khandwa 3 (68 KH-33-1146) Jawahar Kapas 3	1984	9	35.0	23.0	30s	Nimar (MP)
Purnima (NH 239)	1985	10–12	39.5	23.0	30s	Marathwada (MS)
G. Cot 14 (G 2637)	1986	20	36.0	26.5	36s	South and mid. Gujarat
AKH 081 (PKV 081)	1987	20	36.0	24.0	34s	Vidarbha (MS)
LRK 516	1992	30(15R)	36.0	25.0	36s	GS, MS, S. Rajasthan
LRA 5166	1992	20–25	35.0	29.0	34.0	MS, GS, MP, (Vide South Zone)
JK 4	2001	18–20	35.0	27.0	35's	MP
G. Cot 18	2001	15–18	34.0	26.0	35;s	Gujarat
Pratima	2001	15–18	34.0	25.0	35's	Maharashtra

Table 2: (*Contd...*)

Table 2: (*Contd...*)

Variety	*Year of release*	*Seed cotton yield potential (q/ha)*	*GOT (%)*	*M.F.L (mm)*	*Spinning (HSC)*	*Tract/State to which recommended*
		Desi cotton Varieties				
G. Arboretum						
Jyoti	1973	6–7	39.0	24.0	26s	Khandesh (MS)
AKH 4	1975	7	39.0	24.0	20s	Vidarbha (MS)
AKA 5	1981	7	39.0	22.5	20s	Vidarbha (MS)
Eknath (PA 32)	1981	9	39.0	21.5	20s	Marathwada (MS)
Rohini (NA 48)	1984	11	39.0	24.5	26s	Marathwada (MS)
G. Cot 15	1989	10	34.1	21.8	20s	South & middle Gujarat
AKA 8401	1989	10	38.0	27.0	40s	Vidarbha (MS)
AKA 7	2001	14–16	35.0	25.0	30's	Marathwada (MS)
G. Herbaceum						
Sujay (3948)	1972	7	37.5	23.5	34s	S. Gujarat
G.Cot 11 (1449)	1979	10	39.0	23.5	24s	Gujarat
G.Cot 13	1981	10	39.0	23.0	30s	Wagad zone (GS)

Source: Anon (2004), Anon (1995)

species is presented in Table 4.2. The most notable releases of the south zone are Suvin, MCU 5, MCU 7, MCU 5 VT, LRA 5166, Abadhita, DB 3–12, Savita, Varalaxmi, DCH 32, HB 224 and DDH 2. LRA 5166, MCU 5 and DCH 32 have made a great impact. LRA 5166 has also made a significant impact in the central zone especially Maharashtra state. LRK 516 developed at CICR, Regional Station, Coimbatore has been released for the central zone states of Maharashtra, Gujarat and South Rajasthan during 1992.

The spinning spectrum of the cotton varieties released in the south zone ranges from 20s-120s counts. Duration ranges from 145 days (rice fallows cotton) 210 days (Suvin).

Table 3: Improved varieties released in the south zone

Variety/ hybrid	*Year of release*	*Potential yield of seed cotton (q/ha)*	*GOT (%)*	*MFL (mm)*	*Spinn-ing (HSWC)*	*Recomm-ended for*
***Barbadense varieties* (Egyptian cotton)**						
Sujata	1969	15	30.0	30.5	100s	T.N. and A.P. (irrigated pockets)
Suvin (62/17)	1974	20	29.0	35.5	120s	—do—
TNB 1	1981	30	30.0	31.0	80s	TN (irrigated)
***Hirsutum varieties* (American cotton)**						
MCU 5	1968	20–30	34.0	29.5	60s	TN and AP (irrigated and rainfed)
Mysore Vijaya	1968	20	34.0	30.0	40s	K.S. (Ghataprabha region)
Krishna (AC122)	1968	12	35.0	25.0	36s	AP (rice fallows)
MCU 6 (Bharathi)	1970	8	36.0	24.0	36s	TN (Karungannitract)
Mahalaxmi (1301 DD)	1972	7	36.0	23.5	34s	AP (rainfed)
MCU 7	1972	14	35.0	23.5	34s	TN (rice fallows)
Bhagya (GS 23)	1972	20	36.0	24.0	36s	K.S. (low rainfall areas)
MCU 8	1974	20	35.0	28.0	60s	TN (irrigated winter)
Suman (CP 15/2)	1976	20	36.0	24.5	40s	TN (irrigated) Kerala
Amaravati (AV 1661)	1978	20	35.0	27.0	50s	AP (Nagarju nasagar area)
MCU 9	1978	20	36.0	29.0	60s	TN (winter Canbodia tract)
Sangam (V 14)	1978	12	34.0	25.0	40s	AP (rice fallows)
Sharada (CPD8-1)	1980	12	38.5	28.0	40s	K.S. (rainfed)
DS 59 (Saubhagya)	1981	25	36.0	28.0	40s	K.S. (Tungabhadra project)

Table 3: (*Contd...*)

Table 3: *(Contd...)*

Variety/ hybrid	*Year of release*	*Potential yield of seed cotton (q/ha)*	*GOT (%)*	*MFL (mm)*	*Spinning (HSWC)*	*Recommended for*
MCU 5 VT	1982	30	34.0	29.0	60s	TN (V. wilt prone areas)
LRA 5166	1982	28 (10R)	35.0	24.5	34s	TN, AP, MS (Also GS and MP)
MCU 10	1982	8	36.0	25.0	40s	TN (rainfed)
NA 247 (Vijaya-laxmi)	1982	8	35.0	24.0	40s	AP (Rayala-seema)
Supriya (CP 1998 F)	1984	20	37.0	26.0	50s	TN, AP (irrigated)
Kanchana (LPS 141)	1987	28	34.5	25.0	40s	South zone
Abadhita	1988	25	37.0	26.5	50s	K.S. (rainfed)
MCU 11	1988	25	35.0	26.5	50s	TN (winter Cambodia)
JK 119 (Arunabha)	1988	25	40.0	27.5	50s	K.S. (Tunga-bhadra tract)
NA 920 (Priya)	1988	25	38.0	24.5	40s	AP (Rayala seema)
SVPR 1	1991	16	35.0	23.5	40s	TN (summer irrigated)
Sahana	2001	18-20	34.0	27.0	35's	Karnataka
Sumangla	2001	18-20	34.0	28.0	40's	TN
***Arboreum varieties* (Desi cotton)**						
K 8	1971	6	36.0	25.5	36s	TN (Karun-ganni tract)
Saraswati (12009)	1978	6	36.0	24.5	36s	AP (Godavari tract)
K 9	1978	6	36.0	25.0	30s	TN (Karun-ganni tract)
Srisailam (1512)	1976	6	37.0	22.0	26s	AP (Rayala-seema)
Mahanandi (355 E6)	1978	6	31.0	22.0	26s	AP (Rayala seema)
K 10	1985	7	35.0	24.5	30s	TN (Karun-ganni tract)

Table 3: *(Contd...)*

Table 3: *(Contd...)*

Variety/ hybrid	*Year of release*	*Potential yield of seed cotton (q/ha)*	*GOT (%)*	*MFL (mm)*	*Spinning (HSWC)*	*Recommended for*
***Herbaceum varieties* (Desi cotton)**						
Raichur 51	1968	5	31.0	20.5	24s	K.S. (Raichur tract)
DB 3-12 (Ajantha)	1983	8	34.0	21.5	24s	K.S. (Herbaceum tract)

Source: Anon (2004), Anon (1995)

MCU 5 development at the Tamil Nadu Agricultural University and released for cultivation in 1968, represented the first breakthrough in improving quality; this variety belonging to the Superior Long Staple group recorded high yields combined with wide adaptability and spread to other states like Andhra Pradesh and Maharashtra. Significant breakthrough in quality was achieved with the release in 1969 of Sujata cotton, capable of being spun to 100s count, and the released in 1974 of Suvin cotton, capable of being spun to 120s count, and considered as one of the best quality cottons in the world. Both Sujata and Suvin were developed at the CICR Regional station, Coimbatore.

References

Anonymous (1990). Cotton Scenarion in India, CICR, Nagpur. p. 108.

Anonymous (1992). Silver Jubilee Commemoration. Resume of Research in Madhya Pradesh. p. 49.

Anonymous (1994). Seven decades of cotton research in Madhya Pradesh. p. 217.

Anonymous (1995). Hybrid Cotton in India: A Success Story. p. 37.

Anonymous (1992). All India Coordinated Cotton Improvement Project Achievement. Silver Jubilee (1967–1992). Central Institute for Cotton Research. Central Institute for Cotton Research, Nagpur. p. 176.

Anonymous (1999–2000). Project Coordinator Report. p. 186.

Anonymous (2004). Achievements in Citton Research. All India Coordinated Cotton Improvement Project. p. 104.

Anon (2004). Achievements in cotton research, IAICCIP, ICAR, p. 103.

3

Hybrid Cotton Development

Since the time India attained independence, research efforts in cotton were intensified with the 3 major objectives of increasing the total production, raising productivity per hectare and to produce adequate long and extra-long staple cottons to meet the demands of the large Indian Textile Industry and ever – growing population. Heterosis breeding was adopted as a major genetic approach for achieving these objectives through the development of high – yield potential and high – quality hybrid cottons particularly in central and southern India.

The determined and dedicated approach for the commercial exploitation of hybrid vigour particularly in Surat, Indore, Dharwad and Coimbatore research centres led to the development of pioneering hybrid cottons, 'H 4' in Surat and 'Varalaxmi' in Dharwad in late sixties. Further momentum was gained with the release of 'JKHY 1' from Indore and 'H 6' from Surat in the latter part of the seventies and India became the pioneer in ushering a hybrid cotton era in the last 2 decades.

The success in hybrid cotton development was achieved after long years of research and challenging decision to produce hybrid seed by hand emasculation and pollination for commercial cultivation. In the year 1965–66 India's first intra-*hirsutum* hybrid H-4 was released (Fig. 1) involving well adapted popular Indo-American extra long staple variety Gujarat-67 (G-67) as female parent and an-exotic *G. hirsutum* variety American Nectariless as male parent. The cultivation of hybrid cotton in Madhya Pradesh, Maharashtra was introduced in 1969–70 basically for irrigated area using H-4 hybrid. By about the same time hybrid seed production programme of H-4 was undertaken and the scientists of the All India Coordinated Cotton Improvement Project both at Khandwa and Indore worked together with the State Department of Agriculture personnel as extension and problem solving teams. Later in 1972–73, inter-specific hybrid Varalaxmi appeared on the scene from Karnataka (Anonymous, 1995). This marked the complete turn of events making both inter-specific and intra-specific hybrid cotton a reality of commercial hybrid production in India as well as in the state.

Simultaneously with the establishment of H-4 and the hybrid seed production technology in the early seventies, workers had also toned up with the thought that hybrid had come to stay in the country and added one more element *i.e.* resistance hybrid breeding programme. Concurrently, endeavors in Madhya Pradesh, succeeded in obtaining crossed seed for top cross evaluation tests in 1971–72. Initially Khandwa-2 was crossed with several pollinators. The choice of female parents basically required to be tolerant/resistant to the early season sucking pest particularly Jassid. Beside this, the female Khandwa-2 was also known to be high yielder under stress, high ginner, superior high medium staple type with good strength.Several cross combinations were tested in comparison with H-4.

The hope were pinned on two specific combinations- Khandwa-2 (*G. hirsutum* race latifolium) x Reba-B-50 (*G. hirsutum* race punctatum) and Khandwa-2 x ERB-4492 (*G. barbadense*). The choice of this combination was just not an accident but was adequately planned. Khandwa-2, a jassid tolerant early maturing medium staple variety with wide adaptation as such was most suitable variety to be used as the female parent, whereas, Reba-B-50 but for susceptibility to sucking pest, and restricted adaptation also carried a dominant gene (Bgl) for resistance against *Xanthomonas malvacearum* which was tightly linked to the dominant gene(s) for yellow pollen colour which was also a mark of hybridity in this combination. This F-1 combination was released in 1976 as Jawahar Kapas Hybrid-1 or JKHy-1 or JKH-1. Another important development was the release of first inter-specific hybrid JKHy-11 (K-2 x ERB-4492). This hybrid was known for its extra long fibre quality. One of the remarkable features of the work attended to hybrid development programme was the maintenance breeding programme in order to continuously maintain and improve the performance of the parental line.

After almost twenty years, another successful combination involving Vikram x Reba-B-50 was developed and released as F-1 hybrid JKHy-2. The hybrid showed superior performance over JKHy-1 (Anonymous, 1994).This hybrid is early maturing suitable for double cropping system. It performed well under irrigated and rain-fed agro-ecological situation of M.P. The hybrid has very good fibre properties suitable for modern spinning machine. This hybrid is also tolerant to bollworm due its early nature, which escape the peak bollworm infestation period.

Another important event in the heterosis breeding is development of intra-hirsutum hybrid JKHy-3. It was released for the entire cotton growing area of the state of M. P. It is suitable for both irrigated and rain-fed situation. It has good plant type among the different hybrids. It is 100–125 cm tall, bushy with short fruiting inter node. It is early maturing (160 days) as compared to JKHy-1 and tolerant to sucking pest.

SPECTRUM OF HYBRID COTTON

The spectrum of hybrid cottons developed in India is very wide and unique in many aspects and till today this achievement has not been surpassed or realized by any other country. Hybrids have been developed in intraspecific and interspecific categories at tetraploid and diploid levels.

Intra-hirsutum hybrids. 'H4', 'H6', 'JKHY 1', 'PKV Hy 2' ('AHH 468'), 'Lam Hybrid 1', 'G. Cot. Hy 8', 'MECH 1', 'MECH 4', 'Ankur 15', 'Ankur 51', 'NHH 44', 'Savitha' ('HH712'), 'Somnath', other MECH series, 'CAHH 468', 'Suguna' ('CPH 2'), 'CICRHH 1' 'JKHY 2" and 'JKHY 3'.

Intraspecific hirsutum x barbadense hybrids. 'Varalaxmi', 'CBS 156', 'RHR 253' ('Savitri'), 'NHB 1', 'KCH 1', 'JKHY 11', 'DCH 32' ('Jayalaxmi'), 'HB 224', 'NHB 12', other MECH series, 'TCHB 213'.

Diploid desi cotton hybrids 'Ah 71', 'DDh 2', 'G. Cot. DH 7', 'G.Cot. DH 9'.

In spite of the fact that a large number of hybrids have been developed, the popular hybrids as on date are 'H 4', 'H 6', 'JKHY 1', 'PKV Hy2', 'NHH 44', 'G. Cot. Hy 8', 'MECH 1', MECH 4 'Varalaxmi' and 'DCH 32', occupying a significant area. The potential characteristics for these popular and promising hybrids are presented in Table 1.

Table 1: Potential characteristics of popular and promising hybrids

Hybrids	***Year of release***	***Yield of seed cotton (q/ha)***		***Staple class***	***Spinning potential (HSC)***	***States widely growing/ recommended***
		Average	***Highest***			
Intra-hirsutum						
'H 4'	1970	30	60**	Long	60s	Gujarat, Maharashtra, Madhya Pradesh, Andhra Pradesh
'JKHY 1'	1976	25	50*	Long	50s	Madhya Pradesh, Andhra Pradesh
'H 6'	1979	30	60	Long	60s	Gujarat, Maharashtra, Madhya Pradesh, Andhra Pradesh
'PKV Hy 2'('AHH 468')	1981	25	35	Superior -medium	40s	Maharashtra
'PKV Hy 4('AHH 468')	2001	25	40	Superior -medium	40s	Maharashtra
'MECH 1'	1983	25	60	Extra-long	60s	Madhya Pradesh, Andhra Pradesh
'NHH 44'	1983	25	50	Superior -medium	40s	Maharashtra
'Savitha'	1987	30	60	Extra -long	60s	Tamil Nadu, Andhra Pradesh
'G.cot. Hy 8'	1988	35	60	Superior-medium	40s	Gujarat
'MECH 8'	1990	35	60	Long	60s	Maharashtra
'CICR HH 1'	1990	30	60	Superior-medium	40s	Maharashtra
JKHY 2	1998	25	40	Long	50s	Madhya Pradesh,
JKHY 3	2002	25	35	Superior medium	40s	Madhya Pradesh
Phule 492	2002	25	40	Superior	40s	Maharashtra

Table 1: (*Contd...*)

Table 1: *(Contd...)*

Hybrids	*Year of release*	*Yield of seed cotton (q/ha)*		*Staple class*	*Spinning potential (HSC)*	*States widely growing/ recommended*
		Average	*Highest*			
Hirsutum × barbadense						
'Varalaxmi'	1972	25	60	Extra-long	80s	Karnataka, Andhra Pradesh, Tamil Nadu
'DCH 32' (Jayalaxmi)	1981	30	60	Extra -long	100s	Karnataka, Andhra Pradesh, Tamil Nadu, Maharashtra, Gujarat
'NHB 12'	1989	30	60	Extra -long	80s	Maharashtra
'HB 224'	1989	25	60	Extra -long	80s	Tamil Nadu, Andhra Pradesh, Karnataka
'TCHB 213'	1989	25	55	Extra -long	80s	Tamil Nadu
Desl cotton hybrids						
'G. Cot DH 7'	1985	15R	25R	Medium	20s	Gujarat
'DDH 2'	1987	10R	15R	Medium	20s	Karnataka
'G. Cot DH 9'	1988	15	25	Long	50s	Gujarat

Source: Anon (2004), Anon (1995)

Cotton hybrids are available in medium, superior-medium, long and extra-long staple categories with spinnability range of 20s to 100s counts. The hybrid cottons have proved their export potential and suitability for easy care treatments and blending. Early maturity with 150 to 170 days duration is found in certain hybrids like 'H 6', 'G. Cot. DH 8', 'Savitha', 'PKV Hy 2', JKHy 3 and CICR HH 1'.

Popular Cotton Hybrids in Cultivation

***JKHy-1*:** This hybrid was released in 1976–77 and is still very popular among the farmers. This single hybrid covers more than 75,000-

hectare area of the state of M.P. This hybrid is also grown in Andhra Pradesh. This hybrid is also widely grown under rainfed condition. Its average yield potential under irrigated condition is 25–35 quintal/ ha and under rainfed condition is 20–25 quintal/ha. It requires less number of sprays for the control of jassid. With the introduction of *Bt* hybrids, there is a great demand from the farmers to convert this hybrid into *Bt* background.

H-8: This is another important public sector hybrid which is very popular and grown in large area. It is early maturing, more tolerant to jassid and permits double cropping in irrigated area. The negative point associated with this hybrid is its fibre quality (poor length and strength). Due to its earliness the problem of wilting some time occurs due to sudden moisture depletion under shallow soil condition. This hybrid is in cultivation in M.P. and Gujarat.

***NHH-44*:** This is *intra-hirsutum* hybrid and mostly cultivated in deep soils under irrigated condition. Its area is gradually declining because of several private sector hybrids. Tolerance to sucking pest is also very high. The yield potential is at par with JKHy-1. This hybrid is not suitable under rainfed condition.

***DCH-32*:** This inter-specific hybrid is in cultivation since 1970–71 in Madhya Pradesh, Karnataka and Tamilnadu. This hybrid is known for its excellent fibre quality. Its fibre has very low motes content. The plants exhibit synchronized flowering and fruiting habits. The hybrid is tolerant to blackarm and red leaf blight besides being amenable for efficient pest management due to its less leafy nature. In the early stage of plant growth up to 60 days it requires 2–3 sprays for the control of sucking pest particularly jassid. This hybrid is suitable under irrigated condition with deep soil. The average yield potential of this hybrid under irrigated condition is 18–20 q/ha

JKHy-2: This is another intra-hirsutum hybrid developed by the scientists of Madhya Pradesh. This is an early maturing hybrid (160 days) suitable for double cropping system (cotton + wheat one year). This hybrid gradually becomes popular among the farmer community. The hybrid is suitable for irrigated and rainfed condition for entire M.P. In Chhindwara Distt. this hybrid have shown excellent performance and most of the farmers prefer to grow this hybrid. The average yield potential of this hybrid under irrigated condition is 25–30 q/ha. The fibre quality is very good and suitable for modern spinning machines. The fibre length of this hybrid is 26–27 mm with 23 g/tex strength and spinnable upto 50 counts.

***Jawahar Akash (JKHy-3)*:** This is another important public sector intra-hirsutum hybrid developed by the cotton breeders at Khandwa. The important feature of this hybrid is its wider adaptability and suitability under both irrigated and rainfed condition. It will replace the JKHy-1. The hybrid is tolerant to sucking pests and more resistant to black arm disease. The average yield of this hybrid is 25–30 quintal/ha under irrigated condition and 20–22 quintal/ha under rainfed condition. The hybrid has good plant type. It grows about 110–120 cm. tall and open type.

***H-10*:** This hybrid is in cultivation in the Gujarat, Maharashtra and Madhya Pradesh This hybrid is suitable for irrigated area with deep soils. The yield potential is comparable with other ruling hybrids. It is having good plant type with big boll size. The average yield potential of this hybrid is 30–35 q/ha.

Impact on Production and Quality

Encouraged by the success and popularity of these hybrid cottons, the tempo of hybrid cotton research was not merely maintained, but various new possibilities for further strengthening were vigorously pursued leading to the release of several more hybrid cottons during the eighties. Starting from a mere 23,000 hectares in 1970–71, the area under hybrid cottons rose to nearly 2 million hectares representing 27% of the national cotton area in 1989–90, which is a record production year in the cotton history of India, with 12.8 million bales, of which, nearly 45% contribution came from hybrid. The contribution of hybrid cottons to the attainment of self-sufficiency and exportable surplus in long and extra-long staple categories and increased production and productivity per hectare has now been well recognized, A major proportion of long and extra-long staple cottons came from the hybrid cottons besides some medium staple.

The hybrid cottons have played a very crucial role in the attainment of self-sufficiency and exportable surplus in long and extra-long staple categories and have also increased production and productivity.

Development of CMS Hybrids

Use of male sterility in seed production has been an important objective. Except a few hybrids like ', 'CAHH 468' and certain MECH hybrids, all the remaining hybrids for commercial cultivation are produced through conventional Doak's method of the hand-emasculation and hand-pollination. Some commercially popular conventional hybrids like 'H 4', 'H 6', 'Varalaxmi'and 'JKHY 1' were reconstituted on the male sterile system; but not adopted commercially so far.

Studies with genetic male sterility (GMS) *vis-à-vis* cytoplasmic male sterility (CMS) along with restorers have indicated the potential for achieving higher heterosis levels in GMS (Gregg source) than CMS, notwithstanding other problems in seed production. This is attributed to wider chance for parental choice on the male side. *G. harknessii* continues to be the only source of CMS besides limitation of effective and usable restorer sources. However, the scope of using CMS may improve with availability of more number of converted male sterile and restorer lines and thereby identifying higher heterotic combinations. Although use of male sterility enables us to dispense with hand-emasculation, the greater benefit could occur when hand-pollination in hybrid cotton seed production is also replaced by cheaper methods. Use of male sterility alone can reduce the cost of seed production by 50–60%. Research on the possibility of use of honeybees for pollination is in progress in some centers in the USA as well as in India. Use of male sterility along with effective pollination methods can revolutionize hybrid cotton seed production and it would be possible, thus to meet the massive requirement of over 5,000 tonnes of certified hybrid seeds. With this idea in view, research on GMS systems and pollination techniques has been intensified for reducing cost of hybrid seed production so that even poor farmers would be in a position to adopt cotton hybrids. The problem of linkage of CMS hybrids to poor fibre property is the major hurdle in promoting the CMS technology in cotton.

A special project for the promotion of hybrid cotton research has been implemented by the ICAR from 1989–90 in 11 centres in India for evolving suitable hybrids for the different cotton zones of the country especially for northern zone comprising Punjab, Haryana and Rajasthan. The northern zone could not participate in the hybrid cotton revolution probably due to non-availability of suitable potential hybrids which could out yield the already high-yielding cotton varieties grown there. The existing high potential hybrids of southern and central zones are relatively of longer duration than the length of cotton season available in the north zone. Other chief objectives for the future efforts include the development of suitable desi hybrids for areas which are frequently exposed to ecological stress such as drought, development of short-duration tetraploid hybrids with superior yield potential and pest tolerance,

Hybrid cottons have become very popular in India making it the only country in the world to achieve this distinction. Hybrid cottons have contributed for the increased productivity by at least 50% over varieties, higher yield from rainfed areas, higher returns to growers, a flourishing seed industry, high rural employment especially for farm-women, enlarged qualitative spectrum of Indian cottons, high competitive power of cotton with synthetics, India's leadership in the production of long and extra-long staple cotton in the world, exportable surplus and high utilization of cottonseed for edible oil and protein. The potential of hybrid in the north zone, which is fully irrigated has know been exploited to some extent with the spread of *Bt* cotton.

References

Anonymous (1990). Cotton Scenarion in India, CICR, Nagpur. p. 108.

Anonymous (1992). Silver Jubilee Commemoration. Resume of Research in Madhya Pradesh. p. 49.

Anonymous (1994). Seven decades of cotton research in Madhya Pradesh. p. 217.

Anonymous (1995). Hybrid Cotton in India: A Success Story. p. 37.

Anonymous (1992). All India Coordinated Cotton Improvement Project Achievement. Silver Jubilee (1967–1992). Central Institute for Cotton Research. Central Institute for Cotton Research, Nagpur. p. 176.

Anonymous (1999–2000). Project Coordinator Report. p. 186.

Anonymous (2004). Achievements in Citton Research. All India Coordinated Cotton Improvement Project. p. 104.

4

Agronomic Practices Followed for Cotton

G. arboreum

In general the *G. arboreum* cotton varieties are suitable for rainfed condition with diversified soil and climate. The newly developed *arboreum* varieties are also suitable for cultivation in partial irrigated condition giving 20–25% additional yield.

Suitable soil type: Cotton is grown in different types of soil *viz.*, sandy, sandy loam, clay loam, medium to deep black soil. However, medium to deep soils are the most suitable type for cotton.

Preparation of land: Deep ploughing once in three years is recommended which help in rapid root and shoot growth. It also ensures sufficient moisture in the root zone and reduced competition with objectionable weeds. By deep tillage hard pan on the upper surface of the soil is also broken which helps the root growth and free movement of water. Deep ploughing is followed by harrowing twice which helps the land to be broken in to small pieces and finally leveling is done by a leveler. The stubbles of the previous crops are also removed before sowing.

Sowing time: Sowing of cotton is usually done with the onset of monsoon (end of June). In fact, dry sowing has proved to be an excellent method of sowing. This method of sowing is mostly practiced in central cotton growing zone (M P and Maharashtra). During the last few years early sowing is being (mid May) adopted where facilities for drip irrigation has been established and water resources are available.

***Method of sowing*:** The sowing is done by square planting with 60 × 60 cm. row to row and plant to plant spacing in light soils under rainfed cultivation. Under irrigated condition the spacing followed is usually 60 × 90 cm. Two to three seeds per hill are dibbled in cross points made by kolpa nearly 5 to 6 cm below the soil. This system is also known as *chaufully method* and is very popular in central zone. In the last few years after introduction of *Bt* hybrids the spacing followed is 90 × 90 cm.

Another system of planting seed is drilling in which uniform spacing between row is maintained *i.e.* 60 cm row to row spacing but with in the row plants are not regularly spaced. This system is not in practice now.

***Delinting of cotton seed*:** Fuzzy seed are difficult to grade the seed into mature, immature and damaged or cut seeds. Therefore delinting of seed by treating with concentrated sulphuric acid is recommended. For one kg seed 200 g acid is required. For acid delinting sieved and dry seed is put in earthen/plastic containers and measured quantity of acid is poured slowly on the seed and stirred with wooden handle so that all the seed get uniform coating of acid. After 10–15 minutes the seed become blackish due to burning of fuzz and thereafter seed is washed with water 2–3 times. Later on, the seed is washed with lime water to neutralize the acid if any. At final washing with water, the seed flowing on water out are separated immature and undeveloped seed. Finally the seed are separated and dried. Acid delinted seed has many advantage particularly infection of fungal and bacterial diseases is also minimized.

***Seed treatment*:** The seed should be treated with fungicide @ 2–3 g/kg seed to avoid incidence of seed born diseases like BLB and root rot. Seed treatment with bio–flab in combination with 50% nitrogen was found more remunerative.

***Seed rate*:** In cotton there is almost 100% seed replacement every season. The cotton seed which is used for sowing should be genetically pure, viable, vigorous and of high standard. It should be free from any disease infection and admixture.

The seed rate of *G. arboreum* varieties depend upon variety, soil type and growth behaviour. The seed with the spacing of 60 × 60 cm or seed rate of 4–5 kg/ha and with the spacing of 60 × 45 cm or 45 ×

45 cm of 5–7 kg/ha is recommended. The quantity of seed is reduced by 25% if seed is acid delinted.

***Thinning/gap filling*:** Optimum plant population is a basic requirement for getting optimum yield of cotton. In dry sowing due to erratic rainfall and insufficient moisture the seed do not emerge. These gaps resulted into poor crop stand. It is essential to fill up these gaps within shortest period of time.

For proper growth and development of the plant one or two seedlings are to be maintained. The excess should be removed with in 20 days from the date of emergence. The thinning should be preferably done in raining condition. While thinning, due care is to be taken to remove only the diseased, weak and damaged seedlings.

***Weed control*:** Grass and broad leaved seeds emerged with the cotton seedlings which causes heavy losses in cotton yield. Moreover, monocot weeds are more problematic and timely weeding is essentially required for proper growth of the plant. First weeding with in 15 days from the date of emergence is recommended. The weed control is done by one hand weeding followed by continuous inter culture operation by Kolpa and Dora which is most effective and economical. In cotton, none of the weedicides was found to be effective and uneconomical.

***Interculture*:** In rainfed cotton during growing period for effective interculture operation, Dora and Kolpa are used which is easiest and economical way for the control of weed. In desi cotton repeated inter culture in the standing crop till flowering is generally done to control the weed and loosening the top soil after every rainfall which provide proper aeration. It is done 3 to 5 times depending on soil type and growth of the plant. In square planting soil is intercultured across the row to prevent soil erosion during rainy season. The interculture operation is stopped when the bullock movement in the field is not possible due to crop growth. Earthing is done at final inter culture operation with dora which helps in the efficient drainage of excess water and also help for irrigating the field.

***Irrigation and water management*:** Desi cotton varieties are generally recommended under rainfed condition. Variety Jawahar Tapti was found more remuneration with partial irrigated condition. Two irrigation at the time of flowering stage is recommended with minimum and efficient use of water by alternate row method of irrigation. For moisture conservation use of biomulch crop like cowpea, dhaicha, black gram and sunhemp was found most effective.

***Pest and diseases*:** Most of the desi cotton are resistant/tolerant to sucking pests. Only bollworms needs to be managed. Spotted and to some extent American bollworm attack the crop. Beside these, red cotton bug (*Dysdercus koenigii*) and dusky cotton bug (*Oxycaracnus lactus*) also causes damage. For the control of bollworm complex, two spray is recommended by Curacron/Monocrotophos/Quinalphos/ Chloropyriphos. Use of NPV is also recommended. Bird perches can be used for the control of bollworm.

In desi cotton, Root rot (*Rhizoctonia solani*), Fusarium wilt (*Fusarium oxysporum*), Alternaria leaf spot (*Rhizoctonia solani*) and Grey mildew caused by *Ramularia areola* are some of the important diseases in arboreum cotton.

AMERICAN COTTON VARIETIES/HYBRIDS

***Soil type*:** American cotton varieties and hybrids are grown in different type of soil ranged from shallow to medium and deep black. Hybrids are generally recommended in medium to deep black soil.

***Land preparation*:** Pre-planting tillage is very important for proper growth and development for this deep ploughing during summer season is recommended to ensure sufficient moisture and also it control the insect cycle. Deep ploughing is recommended ones in two to three years depending upon preceding crop, soil type and weather factors. Deep ploughing in alternate years has also been helpful in increasing the cotton yield. Deep ploughing followed by harrowing and labeling is done. Before sowing stubbles of previous crop are essentially to be picked and removed.

***Sowing time*:** For irrigated hybrids cotton cultivation, the optimum sowing time is mid May to June while in rainfed condition the seed should be sown with the onset of monsoon. The delay in sowing not only results in extremely low yields but also in poor response to fertilizer, pesticides and other inputs. The dry sowing with the onset of monsoon is generally practiced in MP and Maharashtra. It was found that cotton variety Khandwa–2 and hybrids JKHy–1 can tolerate some delay in sowing both in irrigated condition.

Rainfed cotton cultivation with optimum sowing time has many advantages

- It was found that sowing of cotton at the optimum time achieved maximum plant stand.
- The seedlings are healthy and resist to the vagaries of weather in better way than the late shown cotton.
- Crop sown earlier than the optimum time makes excessive vegetative growth resulting in poor yield.
- Timely sown crop is less susceptible to pest and disease.
- It gives higher yield and good quality of fibre.

***Summer sowing hybrid*:** In part of the Madhya Pradesh and Maharashtra, farmers are taking cotton crop.

***Weed control*:** Cotton crop is highly susceptible to competition from weeds from sowing to about 75 days, when the canopy covers the inter-row space. The problem of weed is more prominent in the cotton crop grown in shallow soil where growths of the weeds are faster than the cotton crop, in the initial stage. Weeds remove the available nutrients from the soil and compete with cotton for nutrient, light and moisture. This competition becomes more detrimental to cotton crop and its productivity in rain-fed area during the periods of drought. Incessant rains in spells make the mechanical weeding impossible particularly in deep black soil. The traditional and still very effective weed control practices followed for hirsutum varieties and hybrids are use of bullock drawn kolpa. One hand weeding in the initial stage of crop followed by 4 weeding by bullock drawn kolpa is commonly practiced.

Application of herbicides like Fluchloralin, Diuron and Fenuron has been evaluated and no single herbicide has been found uniformly effective under all situations *vis-à-vis* manual weeding and weeding with kolpa.

***Nutrient Management*:** Nutrient requirement of crop is decided by the rooting behavior, varieties/hybrids, the fertility status of soil, the potential yield and nutrient management skill. In general, compared to other crops, cotton roots forage deeper (45 cm.) in to the soil and are efficient in availing the native soil nutrients like phosphorus and potassium. The important nutrients, which are applied through chemical fertilizers, are NPK, Sulphur and Zinc.

***Nitrogen*:** Cotton responds to N application under rainfed and irrigated condition. It is supplied in the form of urea. For hirsutum cotton varieties under rainfed condition 60–80 kg N/ha is recommended while 100–120 kg N/ha is recommended for rainfed hybrids. Under irrigated condition doses higher than 120 kg/ha are recommended.

Application of N in 2–3 split (½ at sowing + ¼ at squaring + ¼ at flowering) is more beneficial for hirsutum cotton varieties and hybrids under rainfed condition. Deficiency of nitrogen during flowering and boll formation results in bud and boll shedding. Under irrigated condition 5–6 split is recommended for hybrids.

***Phosphorus*:** Application of phosphorus is reported to increase the yield of seed cotton, boll weight, oil content and tends to bring early maturity of crop. Phosphorus deficiency leads to short and brittle internodes with small and thickened leaves and poor quality of cotton. In P deficient soil application of 40–60 kg P/ha for rainfed hirsutum varieties and 60–80 kg P/ha for irrigated hybrids is recommended. The phosphorus is applied in two splits *i.e.* ½ as basal and ½ after one month.

***Potash*:** Considering the long–term soil fertility, application of 20–30 kg/ha for hirsutum cotton varieties under rainfed condition and 30–40 kg K/ha for hirsutum hybrids is recommended. It was found that when K is applied with N and P, the efficiency of K is increased. It is also applied in two–split–½ dose as basal and ½ dose after one month.

***Foliar application of NPK*:** It was found that foliar application of N in combination with P at the rate of 1.5% at peak flowering stage and 15–20 days later is beneficial. At this time the demand of N is very high and it can be applied through foliar application so that it is immediately available to the plant. At boll development stage phosphorus combination with potash as foliar application at the rate of 1–2% was found to be very effective. It offers strength to cotton to withstand both abiotic and biotic stresses and improve lint quality.

***Placement of N, P and K*:** Placement of NPK fertilizer near the root zone is most effective as it reduces the losses caused by leaching, fixation and weeds. Placing fertilizers 5 cm deep and 5 cm away from the plant either by drilling or by placing in the pockets/ columns is not effective method. Column method comprised of digging 2–3 holes about 20–25 cm deep on the periphery of plant with crowbar and filling them with calculated fertilizer mixture. This method is more effective over ring method to push up the yield by 20–25% irrigated area 30–33% in partially irrigated cotton. The recommendations for quantity of fertilizer application percentage in hybrids and varieties are given in Table 1.

Table 1: Recommendations for quantity of fertilizer application percentage

Duration	***Hybrids***			***Improved varieties***		
At the time of sowing	***Nitrogen***	***Phosphorus***	***Potash***	***Nitrogen***	***Phosphorus***	***Potash***
After 30 days	10%	50%	50%	25%	50%	50%
After 60 days	25%	–	–	25%	–	–
After 90 days	25%	50%	50%	25%	50%	50%
After 120 days	25%	–	–	25%	–	–
After 150 days	15%	–	–	–	–	–

Source: Compiled from Anonymous 1992 a & b

***Other Nutrients*:** Work on micronutrients requirement in cotton indicate that sulphur deficiency prevailed in most of the cotton growing area of M.P. which leads to persistent yellowing of leaves and reddening of petiole. Application of 25 kg $ZnSO_4$ is advocated.

Deficiency Symptoms

All the nutrients are required in a definite proportion by the plant. If any of the nutrients is not available in required quantity, the growth and development uptake and utilization of other nutrients is adversely affected. The deficiency of each nutrient develops a characteristic symptom in the plant that forms a fairly reliable basis of diagnosis. The specific symptoms of nutrient deficiencies are given below.

***Nitrogen*:** Nitrogen deficiency results in retarded and stunted growth, yellow–green and small leaves and few fruiting points. Severe deficiency results in reddening of leaves that become brown and shed prematurely and imperfect opening of bolls known as 'Tiak'. There is small marked decrease in number and size of buds and bolls.

***Phosphorus*:** Symptoms of phosphorus deficiency are not as marked as that of nitrogen. The leaves become small and dark green in colour and stem short and fragile with reduced root growth. The dark green shade of the leaves is followed by diffuse chlorosis and by rust coloured areas near the margins. Severe deficiency reduced the number of bolls and finally the yield.

***Potassium*:** The foliage turns pale green due to yellow white mottling of the leaves accompanied by yellow spots between the veins. These patches rapidly turn necrotic. The tips and margins of the leaves as well as yellow spots become brown and the leaves curl downwards. Finally, the leaves become reddish brown in colour, and shed prematurely. The cotton rust is due to deficiency of potassium that also increases susceptibility to fusarium wilt. There is premature ripening of bolls that frequently do not burst. The fibre is short and immature with poor strength. The leaf blade is hiny in the beginning but later on it losses the appearance. There is marked chlorosis on the apical leaves progressing rapidly to the older ones.

***Calcium*:** The petiole is first curved and later it collapses resulting in marked defoliation. Very few leaves remain attached to the plant. The young buds and leaves become red-dish in colour. Plants are stunted in growth due to death of the terminal buds with poor root development.

***Magnesium*:** A purplish red colour appears in the green veins of the cotton leaves. The lower leaves shed prematurely. Its deficiency results relatively in branched roots and delayed fruiting. The deficiency symptoms first appear on the lower leaves and gradually proceed towards the new leaves.

***Sulphur*:** Sulphur deficiency is often confused with that of nitrogen. The leaves are lemon green in colour and the plant is dwarf in appearance. Very few or no bolls appear on the plant. The vegetative buds in the lower part of the plant start growing.

***Iron*:** Interveinal chlorosis with sharp distinction between green veins and pale green or yellow interveinal spaces. The yellowing of the top leaves in the fine–textured soils appears under water logged conditions. The chlorosis first appears in the upper leaves and after few weeks all the leaves on the top half turn pale green. The leaves are light yellow without spots.

***Boron*:** Its deficiency appears on the terminal buds. Due to die back, the growth is stunted and the plant has many branches. The internodes are short and plant becomes yellow. The young leaves are pale green. The leaves are curled and irregular in shape. There is heavy shedding of buds flowers and bolls.

***Zinc*:** There is retarded growth with short internodes resulting in bushy appearance. The leaves are narrow and small. The younger leaves have pronounced interveinal chlorosis starting from the tip and margin with areas of dead tissues. The veins are darker in colour. The leaves become thick and brittle and margins turn generally upward. The reproductive phase and opening of bolls is delayed.

***Manganese*:** There is discolouration of the leaf blades between the veins which are surrounded by normal green band. Later on necrotic spots appear in the chlorotic areas. The bottom leaves and young bolls are shed prematurely. In severe deficiency, the plant remains in vegetative phase and fails to set fruits.

***Organic and biofertilizer*:** Among the bulky organic manures, which are normally applied to cotton crop, Farm Yard Manure (FYM) and compost are important. Application of 10–12 tonnes/ha of FYM 20 days before sowing is recommended. *In situ* green mannuring, use of sunhemp grown as green manure and incorporated *in situ* is found to improve the productivity and conserve the soil moisture. Cowpea grown as live mulch benefited rainfed cotton cultivation. *Azospirillum* and *Azotobacter* have been found to be useful in supplementary N needs of crop. Soil and seed treatment with there biofertilizer is recommended.

***Irrigation Management*:** The current irrigated area in the state is around 40%. In central zone, the state is having the highest percentage of irrigated area. As the irrigated area increased, the productivity levels went up, quality changed for the better and there was significant increase in cotton production. Alternate furrow or skip-row furrow irrigation is recommended and found most remunerative method.

Drip irrigation combined with furtigation is seen to enhance the yield by 25%. This drip irrigaton system is only suitable in summer cotton and pre monsoon cropping. In summer hybrid cotton cultivation farmers are adopting drip irrigation system. Economy in irrigation water use will aid in enhancing the irrigated cotton area, thereby leading to additional production with better quality.

Table 3: Recommended fungicide for the control of cotton diseases

Disease	***Control measure***
Bacterial blight (*Xanthomonas axonopodis* pv. *malvacearum*)	• Delint the seeds with acid and treat them with fungicides before sowing • Spray strptocycline at 0.01% (100 ppm) + copper oxychloride at 0.2%
Root rot (*Macrophomina phaseolina* and *Rhizoctonia solani*)	• Treat the seeds with carbendazim @ 2g/kg seed or capton @ 3g/kg seed
Alternaria leaf spot (*Alternaria macrospora*) Myrothecium leaf spot (*Myrothecium raridum*)	• Spray mancozeb (0.2%) or Copper oxychloride (0/2%)
Grey mildew (*Ramularia areola*)	• Spray carbendazim (0.1%)
Boll rot (caused by various fungi preceded by attack of boll worm)	• Spray carbendazim along with the recommended insecticides for bollworm

Source: Anonymous, 2003

humus is expected to give better performance. The crop cultivated in shallow soils is likely to encounter stress particularly during peak bolling stage.

Method of planting: Square planting is the normal practice followed for cotton cultivation in the state The seed cotton yield is directly related to the plant population. For maintaining optimum plant population, two seeds should be dibbled at each hill. If one seed is dibbled, arrangement should be made for gap filling by sowing a few seeds (300–500/acre) in polythene bags/ *dona*. Gap filling should be done in dizzling rains by direct placement of the polythene bag (after giving a proper cut) or *dona*.

Spacing: Plant to plant and row to row spacing largely depends on the plant type and the type of soil. Plants with a large canopy require greater space for its optimum growth and hence are needed to be planted with wider spacing. Similarly, luxuriant growth is expected in deep well drained soils and therefore planting should be done at wider spacing in such soils. In medium soils comparatively closer spacing is required to optimize the yields. The seed rate is another factor which depends on the seed index and spacing. It is therefore advised that information on seed index (100 seed weight), plant type and spacing be considered before deciding the quantum of seed to be purchased for a unit area.

***Refugia*:** As per the recommendations of GEAC, 20% of the area or five rows whichever is greater is to be sown by the refugia crop around the main *Bt* crop (Non-*Bt* counterpart of the same hybrid). This is essential for insect resistance management. This refugia crop may be sprayed by appropriate insecticides for controlling the bollworms. It is to be noted that seed of the refugia crop should not be used for gap filling. Mixed sowing of *Bt* and refugia crop should never be done.

***Fertilizer dose and schedule*:** Since *Bt* technology is a crop protection technology, the number of squares and bolls is expected to be more in the *Bt* crop through effective control of bollworms. The requirement of nutrients is therefore expected to be more as compared to the conventional non–*Bt* hybrid. The precise dose of fertilizers will vary from hybrid to hybrid. The exact dose should be decided on the basis of information on the plant type, yield potential, soil type and the number of irrigations that can be given to the crop.

***Foliar application*:** It has been observed in non-*Bt* hybrids that foliar application of N and K helps both in boll development and improving the fibre quality. It is therefore advised that one/two foliar sprays of Urea + Muriate of Potash (1.5%) be applied particularly during peak bolling stage.

A broad range on the dose of fertilizers is given below:

Nutrient	***Dose (kg/acre)***	***Schedule***
Nitrogen as N	60–75	Minimum three splits (1/3; 1/3; 1/3). The first dose as basal and the remaining two two doses at 30 days interval. The entire dose should be completed within 80 days of the crop growth. The dose can be split into more doses depending upon the availability of irrigation and maturity period of the crop.
Phosphorous as P_2O_5	30–40	As basal dose
Potash as K_2O	30–40	As basal dose
Sulphur as S	10	As basal dose
Zn SO_4	10	Soil incorporation. To be applied only in soils where Zn deficiency has been detected.
Mg SO_4	10	Soil incorporation. To be applied only in soils where Mg deficiency has been detected.

Source: Anonymous, (2008)

***Water Management/Irrigation*:** During the last four seasons it has been experienced that *Bt* cotton is highly prone to moisture stress. It is therefore advised that *Bt* cotton should be cultivated under assured irrigated conditions. The first irrigation should be given immediately (within 10 days) of the recession of rainy season. Subsequent irrigations should be given at an interval of 20–25 days. The frequency will again depend on the type of the soil-light soils requiring frequent watering. Ideally three to four irrigations are sufficient for harvesting a good crop.

It is important to note that the problem of sudden drying in cotton is a manifestation of higher atmospheric temperature, moisture stress and higher boll load. Hence water management is not a complete remedy for sudden drying.

***Pest Management*:** *Bt* technology is a crop protection technology and not a yield enhancing technology. The farmer is expected to be benefited by avoiding losses caused by the bollworm complex. The following points about *Bt* cotton will help the farmer to understand the technology and in turn decide the cultivation strategies to derive maximum benefit:

1. *Bt* technology is not effective against sucking pests *viz.* jassids, aphids, thrips and whiteflies. Therefore methods used for controlling sucking pests in conventional (non–*Bt*) hybrids should be used for controlling sucking pests in *Bt* cotton.
2. Insecticidal sprays may be required for controlling bollworms even in *Bt* cotton if the economic threshold levels are crossed.

Thus it is clear that a constant monitoring on the pest population in the field is required to make a decision on when an insecticidal spray is required. Scouting should be done at least twice a week preferably in the morning. Select twenty plants at random in an acre. Count the insects and ascertain that the population is below ETL level. If the ETL level is crossed, the crop should be sprayed with appropriate insecticide. The economic threshold levels of various insects are given below:

***Disease Control*:** The important diseases are Bacterial blight, Myrothecium leaf spot and sometimes Grey Mildew. The following fungicides are effective for controlling these diseases

Table 4: Economic threshold levels of various insects

S. No.	*Pest*	*Economic threshold level*
1.	Jassids	2 nymphs/adults per leaf
2.	Aphids	10% infested plants
3.	Thrips	10 nymphs/adults per leaf
4.	White flies	10 nymphs/adults per leaf
5.	American bollworm, Spotted bollworm and Spodoptera	20 larvae per 20 plants
6.	Pink bollworm	5–10% damaged flowers/bolls

Source: Anonymous (2008)

Table 5: Insecticides effective against cotton pests

Pest	*Insecticide*	*Dose*
Jassids/Aphids	Methyl dematon 25 EC	200–300 ml
	or Imidacloprid 200 SL	40 ml
	or Acetamaprid 20 SP	40 g.
White flies	Triazophos 50 EC or	600–800 ml
	Acetamaprid 20 SP	40 g
American bollworm/	Chloropyriphos 20 EC or	1000–1200 ml
Spotted bollworm	Indoxicarb 15 SC	200 ml
Pink bollworm	Indoxicarb 15 SC or	200 ml
	Cypermethrin 10 EC	320–400 ml
Mites	Dicofol 25 EC	600–800 ml

Source: Anonymous (2008)

Disease	*Fungicide*	*Dose*
Bacterial blight	Streptocycline	1.5g/15 litres of water
Myrothecium leaf spot	Carbendazim	0.15% *i.e.* 1.5g/litre of water
Grey mildew	Carbendazim	0.15% *i.e.* 1.5g/litre of water

Source: Anonymous (2008)

References

Anonymous (1992a). Silver Jubilee Commemoration. Resume of research in Madhya Pradesh. p. 49.

Anonymous (1992b). All India Coordinated Cotton Improvement Project Achievements. Silver Jubilee. (1967–1992): pp. 1–152.

Anonymous (1994). Seven decades of cotton research in Madhya Pradesh. p. 217.

Anonymous (1995). Hybrid Cotton in India: A success story. p. 37.

Anonymous (1992). All India Coordinated Cotton Improvement Project Achievement. Silver Jubilee (1967–1992). Central Institute for Cotton Research, Nagpur. p. 176.

Anonymous (1999–2000). Project Coordinator Report, All India Coordinated Cotton Improvement Project. p. 186.

Anonymous (2003). Nucleus and Breeder Seed Production Manual. NSP. IARI New Delhi. p. 211.

Anonymous (2004). Achievements in cotton research. All India Coordinated Cotton Improvement Project. p. 104.

Anonymous (2008). AICCIP Annual Report, Khandwa. p. 98.

Anonymous (2008). Guidelines for package of practices for cultivation of *Bt* cotton. Compiled by Main Cotton Research Centre, JNKVV, Khandwa.

5

Cotton Seed Production

Providing good quality seed of improved varieties to the farmers through various seed supply systems is a major task in the developing countries, which has a strong agriculture base. Cotton growers of the states used to grow cotton from certified seed, and truthfully label seed purchased from the market or Government agencies. They rarely use their own seeds. Among the four cultivated species, the two Asiatic cottons were grown in more than 95% area, while the introduced upland *G. hirsutum* cotton was grown in less than 5% area at the time of independence. But today, the *G. hirsutum* cottons are grown in 88–90% of the cotton area thanks to the development of HxH hybrid from 1970s and transgenics since 2002. Globally American *G. hirsutum* cottons occupy the largest area contributing to more than 90-92 pr cent of the total production. Among the hybrids in transgenic group, Bunny, Mallika, RCH 2, RCH 20, Brahma, MECH hybrids, Tulsi 4, JKH 99, Ankur 651 and others became popular. Through there are nearly 590–600 private seed companies and some 2000 Bt cotton hybrids released, only some 20–25 command more than 85% in seed distribution.

Seed plays a major role as a basic input in cotton productivity and sustained production. Studies indicated that use of good quality seeds alone could increase production by up to 20 per cent. Good seeds enable the farmers to maintain uniform and vigorous stand of crop, which serve as a catalyst to make other inputs, productive and cost effective.

Varietal Seed Production

The important varieties of *G. hirsutum* and *G. arboretum* category are LRA 5166, LRK 516, Khandwa–2, Khandwa–3, Vikram, JK–4,

Disease and pest management: Integrated pest management technique is advocated based on scouting, economic threshold of key pests, early sowing, clean cultivation, mechanical control of weeds and pests, use of sex pheromones, spraying during effective flowering and fruiting period with synthetic pyrethroids alternated with conventional pesticides like carbaryl and control of carry-over population using only 8–10 sprays in all for irrigated hybrid crops and about 4–6 for rainfed crop. Delinted and pesticide treated seeds are supplied to growers. Seed dressing with fungicides and use of plant antibiotic fungicides are advocated for control of diseases. The

Table 3: Estimated cost of hybrid seed production and hybrid cotton production in optimum conditions (irrigated condition)

	Amount (Rs)/hectare		
Item of expenditure	***Hybrid seed production***	***Cultivation of irrigated***	
		Hybrid	***Straight variety***
Field operation & Management	1800	750	450
Cost of seed	500	300	100
Fertiliser & Plant hormones	3800	2500	2000
Pesticide cost	4000	3000	2300
Labour charges	48000	3000	2200
Irrigation charges	700	250	200
Land revenue etc, other charges	2400	400	400
Total cost	**61200**	**10200**	**7650**
Receipts			
Production and value			
Lint realized from crossed bolls	4000	–	–
Seed in ideal condition	8 q	–	–
Seed cotton yield (optimum condition)	–	30 q	20 q
Value realized by sale of produce (seed/seed cotton including stalks)	94000	18500	10300
Net Profit	**32800**	**8300**	**2650**

Source: Anonymous, 2005

recommended plant protection for cotton seed production is given in Tables 4 and 5.

Beej Swavlamban YOJNA"A Domestic Self Reliant Approach for Quality Seed

Each and every farmer is fully aware of the importance and role of good quality seed as the super most singular factor/input in crop production. Within his resources and sometimes even beyond that, best efforts are made to procure good quality seed. Concerted efforts of Agricultural Research and Development programmes for genetic improvement/up gradation of yield and quality potentials of all agri-

Table 4: Recommended plant protection for cotton seed production

Pest	*Insecticide*	*Quantity of insecticide ha^{-1}*	
Jassids,	Methyl demeton	25 EC	500–750 ml
Aphids,	Imidacloprid	200 SL	100 ml
Thrips	Thiomethoxam	25 WG	100 g
	Acetamiprid	20 SP	100 g
Whitefly	Methyl demeton	25 EC	500–750 ml
	Neem oil + Teepol	–	3.0–3.51 + 500 ml
	Fish oil rosin soap	–	14–15 kg
	Phosalone	35EC	2.5–3.0 litres
Boll worm	Endosulfan	35 EC	2.5–3.0 litres
(Spotted, pink	Chlorpyriphos	20 EC	2.5–3.0 litres
and American	Quinalphos	25 EC	2.5–3.0 litres
bollworm)	Carbaryl	50 WP	1.5–2.5 kg
	Fenvalerate	20 EC	400–500 ml
	Cypermethrin	10 EC	800–1000 ml
	Decamethrin	2.8 EC	600–700 ml
American	Spinosad	48 SC	100 ml
bollworm	Indoxacarb	15 SC	500 ml
	Rimon	10 EC	1000 ml
Spodoptera	Chlorpyriphos	20 EC	1.5–2.0 litres
leaf worm	Quinalphos	25 EC	2.5 litres
	Fenvalerate	20 EC	400–500 ml
	Cypermethrin	25 EC	500 ml
	Decamethrin	2.8 EC	600 ml
Stem weevil	Drenching stem portion on 20th and 35th day with		
	Monocrotophos	40 EC	1.5 ml/liter of water
	Phoaslone	35 EC	2.0 ml/litres of water
Mite	Dicofol	25 EC	1.5–2.0 litres

Source: Anonymous, 2003

Table 5: Recommended fungicide for the control of cotton diseases

Disease	*Control measure*
Bacterial blight (*Xanthomonas axonopodis* pv. *malvacearum*)	• Delint the seeds with acid and treat them with fungicides before sowing. • Spray strptocycline at 0.01% (100ppm) + copper oxychloride at 0.2%.
Root rot (*Macrophomina phaseolina* and *Rhizoctonia solani*)	• Treat the seeds with carbendazim @ 2g/kg seed or capton @ 3g/kg seed.
Alternaria leaf spot (*Alternaria macrospora*) Myrothecium leaf spot (*Myrothecium raridum*)	• Spray mancozeb (0.2%) or Copper oxychloride (0/2%)
Grey Mildew (*Ramularia areola*)	• Spray carbendazim (0.1%)
Boll rot (caused by various fungi preceded by attack of boll worm)	• Spray carbendazim along with the recommended insecticides for bollworm.

Source: Anonymous, 2003

horticultural crops has already been established in overall tremendous progress and food grain self sufficiency.

The seed production has risen to the status of a very large industry both in the private and public sector. Several seed industries with latest technologies and scientific infrastructure are engaged in fulfilling the requirement of various crops keeping in view the agro-climatic and environmental diversities of the country. Despite the enormous endeavors of seed industries and organizations, there is paucity of quality seed. In case of major cereals, pulses, millets and oilseeds, hardly 10–15% of the area is covered under certified seed. The requirement of hybrid seed of commercial crops is never fulfilled. Even the demand of hybrid cotton seed is met to an extent of only 40–50%. However, the demand and supply ratio is further fluctuating due to performance and instability of several research hybrids in different regions. Moreover the socio-economic factors are also largely responsible for the extent of hybrid cotton seed coverage. The cost of hybrid seed which usually is more than Rs. 800/– per kg is beyond the economic capacity of poor farmers and even the middle class average farmers are unable to purchase beyond 25–50% of their requirement. The situation in case of good quality seed of straight varieties of cotton is still worse. More than 90% of the cotton farmers growing general cotton under rain-fed (partly irrigated) conditions have been compelled to purchase the loose seed from open market/ ginnaries. This seed is substandard and mixture of several varieties, F_2s and is not at all seed in true sense. The supply of *Bt* hybrid seeds is entirely from the private sector and the cost of seed to the farmer is almost twice as compared to the cost of conventional hybrids.

The farmers procuring certified seed of cotton varieties for cultivation sell their commercial produce in the market and are unable to retain and recycle the seed because ginning facilities are not available for small lots of cotton. In this way the problem of quality seed in a crop like cotton is more difficult, as compared to other crops, where the farmers practice seed recycling for 2–3 years. Considering the limitations and constraints in the programmes of seed production, storage, transportation and distribution network and management, it is imperative to think of a plausible alternative system so that the poorest of the poor farmer may also become self dependent for his domestic requirement of quality seed at his own with least possible expenses. Based upon the applied importance and aspects of mass selection method for maintenance of good seed at farmer's level, a practical suggestion and idea was floated to the Department of Agriculture during 1994–95 to solve the problem of certified seed of soybean variety. JS–335, which was in great demand and seed requirement could not be met that due to acute short supply and problems of distribution. Just on an experi-mental basis, limited seed of JS–335 soybean was supplied in 5.0 kg small packing kits to hundreds of farmers with mass scale training to raise the crop and multiply the seed with utmost care. This experimental approach was highly successful and thousands of farmers were happy and benefited by producing quality seed of their own. This scheme offered some relief to the government machinery engaged in seed production and distribution system on one hand and it also increased seed replacement rate by best possible multiplication use on the other.

Thus, this idea of distribution of small kits of quality seed for further multiplication to meet out the huge demand at individual farmers level paved the way to launch well planned "Beej Swavlamban Yojna" as a self reliant system at domestic level to solve the seed problem to some extent. Later on, this programme was also organized by the Department of Agriculture to meet out demand of quality seed of *durum* wheat.

This idea was also further implemented to meet out the requirement of hybrid seed in cotton for JKHy–1 and H–8 in the year 1997–98 at the doors of poor farmer (Anonymous, 2005). Under this programme 10–15 thousand pouches containing 10.0 g of female and 5.0 g of male parent seed were supplied to the farmers and training was imparted for production of hybrid seed. This initiative was also successful to a large extent and many farmer could produce 4 to 5 kg of hybrid seed for their requirement by attempting emasculation and pollination through his family members, without any expenditure on labour cost. The farmers were happy and contented when they got 100% pure quality seed of their own with least botherations. In this way even the poorest farmer could get the costly hybrid seed to meet out his requirement, with his own efforts.

This type of approach is more fruitful and effective for multiplication of quality seed of straight varieties of cotton, saturating the village or a group of villages by the farmers themselves, enabling them to utilize the small ginning facilities available at block or Panchayat Level. This is really a community approach programme for quality seed production. This will also help to some extent in making the farmers quality conscious, self dependent while maintaining the seed of straight varieties for 2–3 years successively. As a matter of fact the "Beej Swavlamban Yojna" is a seed production and distribution programme of the farmers, based on the applied and age old tradition of mass selection, which needs to be revived.

Importance of Seed in Cotton Production

The seed quality in case of cotton has very especial and paramount importance for obvious reasons of its genetic implications on the fibre properties desirable for various ultimate end users in the textile industry. The fibre properties are inherent characteristics and the quality parameters for fibre have well defined its quantitative measurements for evaluation and further translation into qualitative terminology. Since the cotton provides the basic raw material called lint or seed fibre, the first and foremost seed quality factor originates from the term "species" *i.e.* the cotton belongs to which of the species (*G. arboreum, herbaceum, hirsutum* or *G. barbadense*). Each of the above mentioned cultivated species of cotton and their races have very well defined characteristic and very specific fibre, lint and technological properties inherent in their seed ever since their origin and evaluation. The natural and artificial selection, plant breeding methods for development of newer varieties, the inter and intraspecies hybridizations, and the hybrids have further transformed the entire spectrum of lint and fibre properties in desired direction not only befitting the needs of spinning/weaving yarn/fabric but suiting the textile world as a whole. All the fibre and lint properties, whether ginning outturn, fibre length, strength, fineness spinning potential, colour, lustre, feel, etc., are inherited characters largely influenced/controlled by genetic factors. Therefore the quality of seed, its pedigree, its purity (genetic as well as physical) has fundamental role in determining the quality of cotton fibre and lint which ultimately would decide its market valuation and price and its ultimate fate in the textile industry. Therefore, before growing the commercial crop of cotton the farmers must give due consideration to its seed if high production and quality have to be obtained simultaneously.

Cost of seed in relation to cost of production

The programme of hybrid seed production of cotton is being carried out on farmers field by the private as well as State Seed Corporations. The foundation seed of male and female parents is supplied by the agencies to the registered farmers, who in turn raise the male and female plants in units of 1 or 2 hectares with best possible plant protection care and produce the hybrid seed through hybridization. The emasculation and pollination is done by manual operation with the help of trained labourers for a period of about 90 days regularly with all precautions as per rules regulation of State Seed Certification agency. The hybrid seed is then put to grow out test (GOT) for genetic purity and thereafter it is processed packed and marketed. This entire programme is very costly where the cost of production comes to app. Rs. 50,000–Rs. 60,000 /- per unit and the cost of hybrid seed ranges between Rs. 400/- to Rs. 500/- per kg. In case of C.M.S./or G.M.S based hybrid seed, the labour cost for emasculation may be reduced. The maximum expenditure in such programme incurred is for plant protection. Thus the hybrid seed production in general is a high input labour oriented costly enterprise. The cost of hybrid seed is also a limiting factor and the cultivation of cotton utilizing hybrid seed is also expensive because of high input and high management cost. Unless and until the hybrid cotton yield levels are two times *i.e.* 100% better than that of straight varieties, the cultivation is not economic and profitable. On the other hand the cost benefit ratio in case of seed production of straight varieties is much better and highly beneficial especially with arboreum cotton varieties. The cost of seed, the seed production and the crop production with *G. arboreum* is least expensive and more profitable, because arboreums are more endurant/tolerant to stress conditions and rain fed situations. Now the arboreum varieties with long fine fibres with spinning potential of 30 counts, high strength and yield potential of 30–35 quintals/ ha are available, which are rather better or comparable to hirsutum hybrids. Considering the demand of textile on one hand and the increasing cost of production, costly inputs, higher use of pesticides, shortage of irrigation water, less profitable yield levels and more risk on the other hand with hybrids, the arboreums are the need of time.

'Beej Swavlamban Yojna "as a tool for self sufficiency of quality seed

The 'Beej Swavlamban Yojna' should be considered as a simple practice of producing good quality seed of varieties from small kits at domestic and community/village level and recycling it for 2–3 years. It is not a substitute of any other programme. It is a door-to-door approach for poor farmers who are deprived of good quality seed due to their socio-economic reasons. The idea is to make them self reliant and self sufficient for their small demands and to harvest good crop with least expenses. This would also attain uniformity in quality as well. There is a need to be popularized this yojna and should be supported in the right spirit and perspective in the overall interest of the farmers and the cotton crop. This approach will help in making available to farmers public bred hybrids and straight varieties.

Table 6: Description of morphological characteristics of cotton varieties

S. No.	*Item*	*Maljari Tapti*	*Jawahar A-51-9*	*Narmada*	*Khandwa –2*	*Vikram*	*Khandwa –3*
1.	Release year	1954	1995	1955	1971	1981	1986
2.	Parentage	Maljari 9 × Jarilla	Maljari × NA-39, NA-48 (white & yellow (flower), NA-32 and G-27	Selection BO-382	CTI type multiline selection KW-66-2096	Selection from IC-1036	A-56-347 (Khandwa) × Acala 44-2
3.	Recommended season	Kharif	Kharif	Kharif	Kharif	Kharif	Kharif
4.	Plant height in cms.	150	150	125	80–100	100–120	80–100
5.	Maturity in days						
	(A) Seed to flowering	60–70 days	50–55	80–90	65–70	55–60	50–60
	(B) Seed to seed	180 days	150	190–240	160	150–160	150
6.	Plant habit	Lanky	Lanky	Lanky	Compact indeterminate	Compact indeterminate	Compact less leafy indeterminate
7.	(A) First flowering days	60 days	55	75	64	45	45
	(B) 50% flowering days	80–90 days	75–80	110	70	70–75	70
8.	Flower colour	Yellow	Yellow	Light yellow	White–Whitish yellow	Cream	Whitish
9.	Petal spot	Present	Present	Absent	Absent	Absent	Absent
10	Pollen colour	Deep yellow	Deep yellow	Creamy	Buff.	Yellow	Yellow
11.	Anther colour	Deep yellow	Deep yellow	Creamy	Buff.	Yellow	Yellow

Table 6: (*Contd...*)

Table 6: *(Contd...)*

S. No.	***Item***	***Maljari Tapti***	***Jawahar A-51-9***	***Narmada***	***Khandwa –2***	***Vikram***	***Khandwa –3***
12.	Stem colour	Purple	Light green with pinkish tinch	Green	Rusty green	Light green	Light green
13.	Node number	6.0	5.5	6.7	5.65	5.5	5.0
14.	Leaf lobing	Narrow okra	Narrow okra	Narrow	Narrow leaved, Deep lobes	Broader	Broader than Khandwa-2
15.	Pubescence of leaf	Hairy	Hairy	Mild pubescent	Highly pubescent long dense hairs on the midrib on the dorsal leaf surface	Mild pubescent	Comparatively less hairy than Khandwa-2
16.	Foliage colour	Dull green	Dull green	Green	Green	Light green	Light green
17.	Resistance to pest & disease	Jassid–R Bollworm–MR	Jassid–R Bollworm–R Bacterial Blight –MR Alternaria Blight–MR	Jassid–MS Bollworm–MS	Jassid–Tolerant Blackarm–MS	Moderately susceptible to jassid and moderately resistant to bollworm	Susceptible to jassid and fairly tolerant to blackarm

Source: Anonymous (2005)

Table 7: Distinguishable morphological characteristics of parents of cotton hybrids

S. No.	*Character*	*American Nectariless*	*G-67*	*G. Cot. 100*	*G.Cot -10*	*K-2 (MB)*	*Reba-(s) B-50*	*ERB-4492*	*DS-28*	*SB (YF)-425*	*Vikram*
1.	Days to flower	40–50	75–120	70–75	67–70	65	60–65	60–65	–	–	50–55
2.	Days to maturity	–	–	240–250	140–150	–	–	–	–	–	160
3.	Plant type	Semi erect	Bushy	Erect growth habit	Open	Bushy spreading indeterminate	Compact relatively determinate	Erect & tall	–	–	Compact indeterminate
4.	Plant height	100–120	150–170	120–150	120–140	90–110	70–80	120–150	120–150	150–160	75–90
5.	Stem colour	–	–	Greenish	–	Green	–	Green	Pinkish	–	Green
6.	Leaf										
	a) Colour	Light green	Green	Green	Dark green	Dull green	Dark green	Dark green	Dark green waxy	Shining green petiole pinkish	Light green
	b) Shape	5 lobes mostly lobes less than half cut	3–5 broad lobes	3–6 lobes & half cut	3–4 lobes	Deep lobed	Shallow lobes	Narrow deeply cut	Medium hairy, petiole with pinkish wash, lamina densely hairy	3 lobes deeply cut	Slightly broad, lobed

Table 7: *(Contd...)*

Table 7: *(Contd...)*

S. No.	*Character*	*American Nectariless*	*G-67*	*G. Cot. 100*	*G.Cot -10*	*K-2 (MB)*	*Reba-(s) B-50*	*ERB-4492*	*DS-28*	*SB (YF)-425*	*Vikram*
	c) Hairy ness	Glabrous	Slightly hairy with leathery feel	–	Hairy	Mild,long hairs on midrib on dorsal surface	Sparsely hairy long hairs	–	–	Glabrous, under surfact sparsely hairy	Slightly hairy
	d) Leaf Nectaries	Absent	Present	–	–	–	–	–	–	–	Present
7.	Flower										
	a) Petal colour	Cream to pale white	Creamy	Pale yellow	Pale yellow	Pale yellow	White	Yellow	Creamy white	Deep yellow	Whitish to creamy yellow
	b) Pollen colour	Buff	Buff	Yellow to pale yellow	Buff	Buff	Deep yellow	Deep yellow	Light yellow	Yellow	creamy to yellow
	c) Petal spot	Absent	Absent	Absent	Absent	Absent	Absent	Present	Absent	Present	Absent
	Boll shape & size	Medium elongated	Big elongated	Medium slightly elongated	Medium round pointed	Big elongated	Medium round pointed	Oblong beaked	Bold roundish	Oblong to roundish	Roundish

Source: Anonymous (2005)

Table 8: Description of morphological characteristics of cotton hybrids

S. No.	*Item*	*H-4*	*H-6*	*H-8*	*JKHy-1*	*JKHy-11*	*DCH-32*	*G. Cot. DH-7*	*JKHy-2*
1.	Year of release	1971	111982	1988	1976	1982	1983	1984	1995
2.	Pedigree	F-G. –37 M-American Nectaryless	F–G. Cot. 100 M–G. Cot 10	F–G. Cot. 10 M– Surat Dwarf	F-Khan. –2 (MB)M –Reba-B –50 (S)	F–Khan. –2 (MB-M – ERB –4492 (S)	F–DS –28 M-SB (YF)–425	F – Sujay M- G-27	F–Vikram M–Reba-B-50 (S)
3.	Plant height (cm)	100–120	120–150	100	100–120	125–175	180–185	150-200	90–125
4.	Maturity in days: A) Seed to flowering	75–80	65–70	40–45	45–55	56–60	50–55	50–55	45–50
	Seed to seed	210–230	190–210	170–190	180–210	220–240	190–200	180–200	170
5.	Flower colour	Creamy white	Yellow	Creamy white	Deepish yellow	Yellow	Creamy yellow	Light red	Creamy yellow
6.	Petal spot	Absent	Absent	Absent	Absent	Present	Medium red spot	Dark red petal spot	Absent
7.	Pollen colour	Buff.	Yellow	Whitish	Deepish yellow	Deep yellow	Yellow	Yellow	Yellow
8.	Stem colour	Green	Green	Rusty green	Green	Green	Green	Light red	Green

Table 8: *(Contd...)*

Table 8: *(Contd...)*

S. No.	*Item*	*H-4*	*H-6*	*H-8*	*JKHy-1*	*JKHy-11*	*DCH-32*	*G.Cot. DH-7*	*JKHy-2*
						surface)		gene)	
9.	Leaf lobing	Broad with light cut	Lobe cut less than half	Broad leaves with deep cut	Deep lobe	Narrow deep cut lobes long dense hairs on the midrib (lower leaf	Large with deep lobes	Lobe cut is more than half leaf nectarines present (marker	Broad leaves with less pointed
10.	Foliage colour	Light green	Light green	Dull green	Dull deep green	Dark green	Dark green	–	Light green
11.	Boll shape	Big roundish	Medium roundish small beak at tip	Roundish	Big broad & pointed at tip.	Roundish beaked	Roundish, dark green pitted, beak not prominent	Boll pitted with red pigmented surface, slightly bended which break at top	Round beaked

Source: Anonymous (2005)

References

Anonymous (2003). Nucleus and Breeder Seed Production Manual. NSP. IARI New Delhi. p. 211.

Anonymous (2005). Cotton seed production manual. Published by Main cotton Research Station, Khandwa, pp. 1–76.

Koutu, G.K. and Mandloi, K.C. (2002). Cotton pp. 247–253. Genetic Improvement of Field Crops. Scientific Publisher, Jodhpurp. p. 335.

6

Bt Cotton in India

The concept of genetic engineering has been exploited to introduce the Bt.gene responsible for controlling cotton against *Helicoverpa, H. armigera, Earias* spp and *Pectinophora gossypiella* that are primarily responsible for restricting the full exploitation of the potential yield. The technology is safe and cost effective and seems to be an effective alternative to the conventional (chemical) method of the insect pest control. Haregreaves (1948) has reported about 1326 insects inhabiting cotton. By and large 230 insects' pest species affect cotton in the world and 46 have been identified as major insects in 32 countries. The major economic hurdle in the cultivation of cotton has been the management of pests of which particular emphasis on the bollworm complex. Looking to the great damage due to these pests, insecticide usage in cotton is very high. Cotton accounts for 22.5% of all crop insecticide sales worldwide (Anonymous, 1995). In India around 55 percent of pesticides used in agriculture are directed towards cotton. In spite of usage of huge amount of pesticide, the control of insects in cotton has been a Psycho-Socio-Economic tension to the farmers. When synthetic pyrethroids pesticides like were introduced in India during 1980, their response was overwhelming. In the process to eliminate the bollworms, these pesticides were overused and the result was the development of resistance in bollworm and resurgence of white flies.

Efforts on breeding for resistance or tolerance to insects in cotton were made during mid 1980's and early 1990. These conventional breeding efforts are aimed for utilizing traits responsible for non-preference, antibiosis and avoidance. A lot of manpower, money and time have been spent for devolving a resistant cultivar. A perfect cultivar resistant to bollworms is still to come. Simultaneous attempts were also being made to exploit the biotechnological tools. The crystal protein gene derived from the soil bacterium *Bacillus thuringiensis* offered great possibilities in the management of bollworms. This gene was transferred to cotton via Agrobacterium with CaMV 35 S. promoter. The Cry protein in transgenic cotton was found to be toxic to bollworms. This was a landmark achievement for controlling the bollworm which is considered to be greatest threat in the economic cultivation of cotton crop (Parlak *et al.,* 1990; 1991; Bartlett, 1995; Barlett *et al.*, 1997; Fitt, 1998).

Cotton has historically been an important commercial crop and has undergone manifold genetic improvement. Given its high chemical dependence, it is not surprising that cotton was one of the earliest crops to be genetically engineered by the industry. It is one of the rapidly adopted transgenic since its commercial debt in 1996. India has a larger area of cotton than any country in the world. Based on the latest estimate, the Directorate of Cotton Development, Ministry of Agriculture reports that 6.3 million farmers planted cotton on 9.4 million hectares in 2008 with an average cotton holding of 1.5 ha (Anonymous, 2009). In 2009, the total area of cotton in India was estimated at 9.6 million hectares approximately 3% higher than in 2008. This increase is in contrast to the 2% decrease in cotton area globally in 2009 versus 2008. The area under *Bt.* cotton has been increasing logarithmically during the last eight years and it is worthwhile to assess whether this increase in production has at least partly been due to the increasing area under *Bt.* cotton.

Comparing the distribution of cotton area in cotton growing states in India in 2008, Maharashtra, the largest cotton-growing state having 3.142 million hectare area, which occupied approximately 34% of India's total cotton area followed by Gujarat (2.354 m h), Andhra Pradesh (1.399 m ha), Madhya Pradesh (0.625 m ha) and remaining in other states of India. Gujarat had 1.30 million farmers, followed by 0.96 million in Andhra Pradesh, 0.45 million in Madhya Pradesh, 0.30 million in Rajasthan, 0.26 million in Haryana, 0.20 million farmers each in Punjab, Karnataka and Tamil Nadu and the balance in other states of India.

The year 1996 is a milestone for genetic engineering in agriculture as insect resistant maize and cotton and herbicide tolerant soybeans were launched for commercial cultivation (Anonymous, 2003).

Transgenic crops incorporate traits such as herbicide tolerance, pest and disease resistance, and product quality improvement and stress tolerance. Over 15 transgenic crops have come into commercial use the world over and transgenic crops covered an estimated area of 58.7 million hectare in 16 countries in 2002 (Anonymous, 2004). In India, *Bt* cotton, the first transgenic crop was approved by Government of India in March 2002 for commercial cultivation for a period of three years. Initially GEAC permitted Monsanto–Mahyco to market their transgenic cotton hybrids namely MECH-12, MECH-162 and MECH-184 in India (*Bt.* cotton), which utilizes protein from *Bacillus thuringiensis.* In the third year the Government also permitted Rasi Seeds to sell their *Bt.* hybrid RCH-2 having the same gene. In 2007, 135 varieties of *Bt* cotton were available based on four events, two bollgard, the one of Ari-Genetic, and a second of *Bt* gene (ISAAA 2010). *Bt* cotton protect the plant from Lepidopteron(American bollworm, Pink bollworm, Spotted bollworm) insects. The insecticidal property of *Bt* cotton differentiates it from other cottons. *Bt* cotton is now being grown in entire cotton growing states of the country. Being the first transgenic crop, it has been subject to extensive monitoring at various stages of its cultivation. It has also been recommended that a series of stakeholders such as government officials, research institutes, agricultural scientists, NGOs be utilized for sensitization and creation of awareness among the farming community.

Adoption of *Bt* Cotton Hybrids in India

Bt cotton, which confers resistance to important insect pests of cotton, was first adopted in India as hybrids in 2002. There were 54,000 farmers which grew approximately 50,000 hectares of officially approved *Bt* cotton hybrids in 2003 while the *Bt* cotton area increased again many fold in 2004 to reach half a million hectares. In 2005, the area of *Bt* cotton in India continued to high scale reaching 1.3 million hectares, which estimated an increase of 160% over 2004. In 2006, the adoption record increases which continued with almost a tripling of the area of *Bt* cotton to 3.8 million hectares. Remarkably in 2006, India's *Bt* cotton area from 3.8 to 6.2 million hectares, to become the largest hectarage of *Bt* cotton in any country in the world. During 2007–08 cotton was grown on 9.44 mha, of which more than 80% was sown to *Bt* cotton. Cotton production in this year was 31.5 million bales (mha) (http://www.cotcorp.gov.in/statistics.asp).

During 2008–09 India is heading towards unprecedented record crop for sixth consecutive year after introduction of *Bt* cotton. There has been reduction in cotton acreage by 2 per cent in 2008–09 and stands at 9.3 million hectares compared to last year (The Cotton Corporation of India, 2008). Indian government has approved 263 *Bt* cotton hybrids till date. The area under cotton during 2008 has come down because of initial deficient rainfall in two major cotton growing states like Maharshtra and Gujarat.

Table 1: State wise *Bt* cotton area from 2002 to 2009 (thousand hectares)

State	*2002*	*2003*	*2004*	*2005*	*2006*	*2007*	*2008*	*2009*
Maharashtra,	25	30	200	607	1,840	2,800	3,130	3,396
Gujarat	10	36	122	150	470	908	1,360	1,682
Andhra Pradesh	8	10	75	280	830	1,090	1,320	1,049
Madhya Pradesh	2	13	80	146	310	500	620	621
Northern region	–	–	–	60	215	682	840	1,243
Karnataka	3	4	18	30	85	145	240	273
Tamil Nadu	2	7	5	27	45	70	90	109
Others	–	–	–	–	5	5	5	8
Total	**50**	**100**	**500**	**1,300**	**3,800**	**6,200**	**7,605**	**8,381**

Source: Cotton Advisory Board 2009; ISAAA, 2009

Bt cotton technology provides new avenues for the Indian textile industry by ensuring availability of good quality raw cotton on regular basis which was a major concern a few years back. In India about 30% of the cotton crop losses were due to pests, diseases and weeds. Of which 10% are due to one major insect pest American bollworm (*Helicoverpa armigera*). For the control of this particular major pest alone about 264 US $ million worth of pesticides were being used in India. Cotton occupies 5% of the total cultivable area, however it consumed 55 per cent of the totally used pesticides in all crop cultivation (Barwale *et al.*, 2004).

COMPARATIVE ADVANTAGES OF *Bt* COTTON OVER NON-*Bt* HYBRIDS IN CENTRAL INDIA

Over the years there has been perceptible erosion in cotton area. This downtrend partly owes to incursions made by pulses and oilseeds crops like pigeonpea and soybean, of late, the decline has been much sharper mainly because of the fact that intensive cultivation of hybrid cotton using repeated and excessive application of pesticides brought about a certain imbalance, consequently the pests went out of control rendering the economics of cotton cultivation far less favorable than earlier days.

The problem has become more complex with the introduction of American cotton (*G. hirsutum*) on a national scale. Earlier most of the cotton grown in India belonged to the desi groups of *G. arboreum* and *G. herbaceum* with *G. hirsutum* occupying a small area. Although the desi cottons were less productive, they were less prone to pest problems. However, the country was in dire need of more cotton and one of the elements in the strategy pursued for attaining this objective was to gradually replace desi types with more productive American cotton. Gradually American cottons came to dominate and along with the intra-species and inter-species hybrids, they now occupy around 85% of the total cotton area.

It was soon realized that under such a dispensation, there would be very serious repercussions in the production and productivity of cotton on account of increasing levels of bollworm damage year after year. The concurrent indiscriminate use of insecticides to mitigate the pest problem on the other hand incurred heavy costs both in terms of money and environmental pollution. The economics of cotton cultivation has therefore become less attractive.

Simultaneously, worldwide, the cotton market is facing major changes. Developing countries will have to exercise due care in adopting modern technologies for forging new opportunities in the changing global market. The Multi Fibre Agreement will phase out from 1st January 2005. The current preferential treatment will cease and market relations will be completely opened. The process of buying will no longer be influenced by the available quota but by competition. The best win of quality, price and delivery time will determine which suppliers get the order and the market options. The phasing out of quantitative restrictions for imports into EU and USA by January 2005 will have a strong impact on global trade in textiles and garments (Basu, 2001; Sreenivasan, 2001). India will have to make structural changes and adjustments to effectively complete in the frontier free global area.

To ameliorate the problem to a certain extent and to make cotton cultivation more attractive and to reinforce the farmers with the capacity to make them more competitive *vis-a-vis* the world market, the Govt. of Madhya Pradesh along with Gujarat, Maharashtra and Andhra Pradesh permitted the cultivation of genetically modified hybrid cotton (*Bt* Cotton) in 2002–2003 especially for the economic management of the bollworm complex.

In *Bt* cotton, due to reduction in the use of broad-spectrum insecticides, the technology has proved eco-friendly and farmer friendly as well. The hazardous effect on surrounding eco-systems and associated insects, wild life and human population is reduced to a great extent as compared to insecticide uses. There is also energy saving coupled with minimization of farm machinery.

Due to application of broad-spectrum insecticides many non-target beneficial insects are also eliminated. It is also proved that *Bt* cotton is toxic to only some lepidopteran insects and are in no way harmful to the non-lepidopteran insects and beneficial fauna (Nakanishi, 1999; Bendict *et al.*, 1996; Lambert *et al.*, 1997).

Apart from increase in cotton productivity, *Bt* cotton has also proved to be eco-friendly due to reduction in plant protection spray. The possibility of reduction 2–3 sprays in *Bt.* cotton was evident in central India resulting in the reduction of cost of cultivation to the tune of Rs. 2000–3000.

Experience on *Bt* Cotton

The *Bt.* hybrid cotton is in cultivation in Madhya Pradesh since 2002–03. In the first year official figures indicate that these hybrids were cultivated in nearly 3500 acres. The following year (03–04), the area sharply rose to more than ten times. In the first year, there were mixed results. There was sudden drying of *Bt.* hybrids in Khargone, Badwani, Dhar and Jhabua districts (Anonymous, 2004). In the second year this problem was negligible. However in third year this problem has again been observed in large scale. The field and plants were critically examined. Information on the use of fertilizers, pesticides and other practices followed was collected. The symptoms and field conditions indicate that the sudden drying in the observed fields is due to new wilt, which is a physiological disorder expressed under the strong influence of the environment. The problem was specifically more in summer sown Bt cotton hybrid MECH–184. The incidence was low in MECH–12 and MECH–162 and also RCH–2. Since the affected hybrids is an early maturing hybrid, the plants were fully loaded with bolls and were thus prone to stress.

During the investigations it was also observed that there were several myths about *Bt.* cotton in the farmers mind. They expected spectacular yields without doing anything. *Bt.* cotton was cultivated both under rain-fed as well as irrigate conditions. *Bt.* cotton was sown even in poor soils. Our experience during the last three seasons through ICMF sponsored advisory project clearly indicate that the farmers have not changed their mindset of excessive insecticidal spray even on *Bt* cotton. Use of synthetic pyrethroids (effective against bollworms) was common on *Bt* cotton. On the other hand we came across some farmers who did not use insecticides even against sucking pests thinking that insecticides are not necessary on *Bt* cotton. It is absolutely clear that insecticides are to be used for controlling sucking pests in *Bt* cotton. Even in case of bollworm complex, insecticides are to be used after proper scouting when the ETL level is crossed. The significance of the refuge crop is also not clear to them. It is one of the essential components for management of insect resistance (Anonymous, 2004).

Survey of *Bt.* Cotton Fields

A survey was conducted by team of scientists to assess the performance of *Bt.* cotton during 2003–04. In this programme, 300 farmers of Khandwa, Burhanpur, Bakaner-Manawar, Dhar-Jhabua, Kukshi-Badwani and Ratlam region of Madhya Pradesh, who had opted for cultivation of *Bt* cotton were contacted (Anonymous, 2004). It was observed that the farmers are crazy for *Bt* cotton and a dramatic increase in area under *Bt.* cotton is expected in the next couple of years. However, most of the farmers are not aware about the recent techniques and know-how for proper cultivation of *Bt* cotton. Several farmers have planted the *Bt* cotton in rain-fed conditions. Majority of the farmers have sown *Bt* cotton crop in summer. The farmers are still indiscriminately spraying insecticides in the *Bt* cotton crop for control of pests. Most of the farmers have sprayed their *Bt*. cotton fields for four times without scouting and surveillance of the fields for insurgence and ET level of insect pests. There are reports of infestation of pink bollworm and new wilt especially in summer sown early maturing *Bt* cotton. The farmers have either knowingly or unknowingly produced the seeds of *Bt* cotton from unauthorized sources.

Biosafety Concerns of *Bt* Cotton

In *Bt*. cotton, insecticidal protein is produced in nature by the bacterium *B. thuringiensis*. This insecticidal protein in the plant begins to breakdown immediately after the plant mature. Thus, it does not accumulate in the soil and leach down to lower strata of the soil. Researches conducted before the application of gene clearly reveal that this gene only accumulates toxin having specific binding sites. Hence, Cry 1 Ac is toxic only to a particular group of insects, mainly the members Lepidoptera and more particularly the cotton boll worm. All other insects including beneficial insects, wildlife, marine life are not affected (Luo *et al.*, 1997).

Bt as an Important Component of IPM

It is true that the bollworm complex is the number one enemy of cotton. However, it is to be borne in mind that *Bt* cotton is not a panacea for all the insect problems. It has the potential to manage bollworms. Here too, chemical control methods will have to be adopted if the bollworms cross the ETL (Fitt *et al.*, 1994; Fitt, 1998; Daly and Greenplate, 1999). The conventional methods adopted by farmers for managing the sucking pests need to be applied in *Bt* cotton as well. An intelligent mix of the technology will be more beneficial for the farmers. The best use of *Bt* cotton would therefore be to use it as a component of the Integrated Pest Management Strategy. The experience during the eight years on the cultivation of *Bt* cotton in the country indicates that even for the management of the bollworm one or two sprays required. Therefore, under the prevailing circumstances. *Bt* cotton cannot be the sole solution for pest management, but *Bt*. cotton genotypes should be used as important component on which integrated pest management has to be built.

Performance of* Bt. *cotton under rain-fed situation in comparison with* arboreum *varieties in Madhya Pradesh

Since the large area under cotton cultivation in the state is rain-fed. Therefore, large scale on farm trails were conducted to assess the performance of conventional long linted *arboreum* varieties , popular *intra-hirsutum* hybrid JKHy–1 in comparison with *Bt.* hybrid under rain-fed condition with a view to recommend suitable cotton variety. The results clearly indicate that *arboretum* varieties Jawahar Tapti, DLSA–17, KWA–25 and PAIG 8/1 are comparable to the hybrids particularly when the C:B ratio is considered which is the single deciding factor while making a choice of a genotype/ hybrid (Anonymous, 2003c) (Tables 2–4). The cost of the *Bt* cotton

Table 2: Overall performance of *Bt* cotton *vs G. arboreum* genotypes on research farm

S. No	*Genotype*	*Seed cotton yield (Kg/has)*	*Ginning percent*	*Plant height (cm)*	*No. of bolls/ plant*	*Cost benefit ratio*
1.	MDL–2463	1069.50	33.33	136.50	26.00	1:3.33
2.	MDL–2593	773.50	33.33	157.50	27.67	1:2.41
3.	MDL–2601	1024.00	35.00	133.83	26.83	1:3.19
4.	PA–405	964.50	31.66	144.50	26.50	1:3.01
5.	PAIG–8/1	1086.00	33.33	159.00	25.17	1:3.38
6.	PAIG–8/3	848.50	33.33	157.50	27.34	1:2.65
7.	KWA–225	984.00	35.00	128.17	25.17	1:3.07
8.	KWA–25	1116.50	32.66	154.17	24.50	1:3.48
9.	DLSA–17	1076.00	33.33	156.00	30.67	1:3.36
10.	PA–255	1033.00	33.33	141.33	26.67	1:3.22
11.	J.Tapti	1122.00	35.00	164.83	31.34	1:3.50
12.	JKHy–1	1033.50	34.16	111.50	30.33	1:2.85
13.	MECH–160 *Bt*	1294.00	33.33	97.84	26.00	1:3.09
	CD(5%)	141.39	4.406	37.28	12.25	
	CV	6.28	5.99	12.07	20.65	

seed alone is a major contributor in increasing the comparative cost of cultivation and works against it, although JKHy–1 and MECH–162 *Bt.* were slightly higher yielder as compared to arboreums. *Bt.* cotton can also be promoted under partially irrigated conditions where one/two irrigations can be given.

Performance of Bt *cotton variety MECH-162 under different agro-ecological situations of Madhya Pradesh*

Under National Agriculture Technology Project at Khandwa, a study was undertaken to identify suitable cotton species for different agro-ecological situations and characterization of identified species for sustainable production under each situation with the ultimate aim to maximize the yield. Six situations were identified on the basis of soil type and rainfall intensity. Seven on

Table 3: Overall performance of *Bt*. cotton *vs G. arboretum* genotypes on farmer' fields

S. No	*Entry code*	*Plant height (cm)*	*Yield (Kg/ha)*	*Yield/ plot (Kg)*	*Ginning percent*	*Cost benefit ratio*
1.	MDL–2463	141.50	1175.00	58.75	32.24	1:3.56
2.	PA–402	139.50	1022.00	51.09	35.12	1:3.09
3.	DLSA–17	135.83	1004.00	42.68	33.70	1:3.04
4.	PA–255	146.50	1163.00	58.13	34.53	1:3.52
5.	Bunny	100.88	1064.00	53.20	34.48	1:2.58
6.	MECH–162 *Bt*	98.33	1246.00	62.32	35.22	1:2.99
	CD 5%	13.72	135.20	6.76	2.68	
	CV	9.07	10.22	10.22	5.51	

Table 4: Fiber quality parameters *Bt* cotton hybrid vs *G. arboreum* genotypes in rain-fed ecosystem at the research farm

S. No	*Entry Code*	*2.5% SL*	*UR (%)*	*Mic value*	*Strength*	*Elongation*	*SFI*
1.	MDL–2463	27.47	45.80	4.32	21.50	6.40	8.66
2.	MDL–2593	28.80	46.00	4.66	22.10	6.30	6.42
3.	MDL–2601	27.10	46.30	4.67	22.50	6.40	8.66
4.	PA–405	29.09	47.50	4.13	21.60	6.20	5.01
5.	PAIG–8/1	28.74	48.00	4.33	23.50	6.40	4.97
6.	PAIG–8/3	27.70	47.80	4.91	21.00	6.20	6.47
7.	KWA–225	28.26	47.50	5.21	20.60	6.30	5.93
8.	KWA–25	26.58	45.90	5.43	21.20	6.30	9.85
9.	DLSA–17	28.45	47.10	4.58	21.10	6.40	6.04
10.	PA–255	29.26	46.70	4.47	21.10	6.30	5.10
11.	J.Tapti	26.95	45.50	4.88	21.40	6.40	9.63
12.	JKHy–1	28.47	47.80	4.26	19.70	6.10	5.37
13.	MECH–162 *Bt*	27.07	46.90	4.13	18.40	6.20	8.22

farm trials were conducted at each situation in which the identified genotype and species of cotton was sown along with MECH–162 *Bt*. The results revealed that the performance of *Bt* cotton in respect of yield and other related parameters is good even under rain-fed situation except in shallow soil with low rainfall situation (Anonymous, 2003d) (Table 5).

Table 5: Seed cotton yield (Kg/ha) of *Bt* cotton crop as compared to other genotypes in different agro-ecological situations under rain-fed conditions

S. No	*Deep soil + High rainfall*	*Deep soil + Low rainfall*	*Medium soil + High rainfall*	*Medium soil + Low rainfall*	*Shallow soil + High rainfall*	*Shallow soil + Low rainfall*
1.	MECH–162 (Bt) (1624.00)	MECH–162 (Bt) (1503.33)	MECH–162 (Bt) (1650.00)	MECH–162 (Bt) (1281.67)	MECH–162 (Bt) (1272.00)	MECH–162 (Bt) (521.67)
2.	Sarvottam (1438.00)	JK–4 (1330.00)	Sarvottam (1340.00)	JK–4 (1121.67)	DCH–32 (974.00)	JK–4 (400.00)
3.	DCH–32 (936.00)	Sarvottam (775.00)	JKHy–1 (884.00)	Sarvottam (641.00)	PA–183 (608.00)	Sarvotam (183.33)
4.	J. Tapti (699.00)	KWH–1 (630.00)	J.Tapi (826.00)	KWH–1 (533.33)	G. Cot. 16 (594.00)	KWA–7 (136.67)
CV	18.6	10.26	18.34	15.58	23.85	13.38
CD	372.83	133.80	296.66	171.52	276.75	51.17

Future line of work

Bt technology is very good, but the success of *Bt* hybrid depends on the choice of hybrid in which the gene has been transferred. The cost of the seed of *Bt* hybrids available in the market is really prohibitive particularly for the marginal farmers. There are three very good options to be followed in the larger interest of the farming community.

1. Conversion of good public and private sector hybrids (like JKHy-1, DCH-32, H-8, NHH-44 and H-10) in to *Bt* hybrids. This hybrid is still in demand even after 20 years.
2. Introduction *Bt* gene in *G. hirsutum* varieties so that the farmers will not be required to purchase seed every year.
3. Introducing *Bt* gene in *G. arboreum* varieties. Arboreums are resistant to sucking pests and additional protection against

bollworm through *Bt* gene will bring down the cost of cultivation significantly. The farmer can use their own seed.

References

Anonymous (1995). Cotton the crop and its pesticides market. *Pesticides News* 30: 1.

Anonymous (2003a). Annual Report AICCIP, Main Cotton Research Station, JNKVV, Khandwa.

Anonymous (2003b). Commercialization of *Bt*. Cotton. Its success and problems in Indian Agriculture. Silver Jubilee Lecture Series Delivered by Dr. B.M. Khadi, Senior Scientist (Cotton), Dharwad. Organized by Indian Society for cotton Improvement on 1st March, 2003.

Anonymous (2003c). Annual Report NATP-7, Main Cotton Research Station, JNKVV, Khandwa. p. 62.

Anonymous (2003d). Annual Report NATP-8, Main Cotton Research Station, JNKVV, Khandwa, p. 54.

Anonymous (2004). Biosafety issues related to transgenic crops. Biotech Consortium India Limited, New Delhi. January 2004.

Anonymous (2008). Cotton Advisory Board, Minutes of the for the cotton year 2007–08 and cotton projection for 2008–09, Office of the Textile Commissioner, Ministry of Textile, Govt. of India.

Anonymous (2009). National cotton Scenario. Ministry of Textile. http/www.cotcoro.gov.in/national_cotton

Basu, A.K. (2001). Impact of WTO on Indian Cotton and Textiles. *National J. PI. Improv.*, 3: 6-10.

Bartlett, A.C. (1995). Resistance of the pink bollworm to *Bt* transgenic cotton. Proceedings of the Beltwide Cotton Conference, National Cotton Council Memphis, T.N.USA. pp. 766–768.

Bartlett, A.C., Dennehy, T.J. and Antilla, L. (1997). An evaluation of resistance to *Bt* toxins in the native populations of the pink bollworm. Proceedings of Beltwide Cotton Conferences, National Cotton Council Memphis, T.N. USA. pp. 885–888.

Benedict, J.H., Sachs, E.S., Altman, D.W., Deaton, W.R. and Kohel, R.J. (1966). Field performance of cottons expressing transgenic Cry 1 A insecticidal proteins for resistance to *Heliothis virescens* and *Helicoverpa zea* (nLepidoptera: Noctuidae). *J. Econ. Entomol.*, 89(1): 230–238.

Daily, J.C. and Fitt, G.P. (1998). Efficacy of *Bt* cotton plants in Australia. What is going on? World Cotton Research Conference-II held at Athens, Greece.

Fitt, G.P. (1998). A future for IPM in Cotton: the challenge of integrating new tools to minimize dependence. Proceedings of World Cotton Research Conference-II held at Athens, Greece. pp. 75–84.

Fitt., G.P. Mares, C.L. and Leewellym, D.L. (1994).Field evaluation and potential ecological impact of transgenic cotton (*Gossypium hirsutum*) in Australia, *Biocontrol Science Technology*. 4: 535–548.

Greenplate, J.T. (1999). Qualification of *Bacillus thurigiensis* insect control protein Cry 1 AC over time in bollguard cotton fruit and terminals. *Journal of Economic Entomology*. 92(6): 1377–1383.

Lambert, A.L., Bradley, J.R. and Vanduyn J.W. (1997). Interaction of *Helicoverpa zea* and *Bt* cotton in North Carolina. *In Proc*. Beltwide cotton Conf., New Orleans, LA 7-10 Jan. 1997. Natil. Cotton Council Am. Memphi, T.N. pp. 870–873.

Luo, K.S., Sangadala S., Masson, L., Mazza, A., Brousseau, R. and Adang, M.J. (1997). The *Heliothis virescens* 170 kDa aminopepitidase functions as " receptor A" by mediating specific *Bacillus thuringiensis* Cry 1 A d-endotoxins binding and pore formation. *Insect Biochem. Mol. Biol.,* 27: 735–743.

Nakanishi, N. (1999). GMO seen as Asia's saviour not Frankenstein Food. Observer of Business and Politics. 22 No. http://observerindia. Com.

Niwanchao, Zhang Zenking, Zhang Baolong and Yingjun, X.N. (1988). Development of transgenic insect protected cotton plants. Proceeding of World Cotton Research Conference-2 held at Athens, Greece. pp. 307–309.

Perlak, F.J., Duff, A.D., Sylvia, L.M. and Fishhoff, D.A. (1991). Modification of coding sequence enhances plant expression of insect control protein genes. *Proc. Natil. Acad. Sci.*, 88: 3324–328.

Perlak, F.J., Deaton, R.W., Armstrong, T.A., Fuchs, R.L., Sims, S.R., Greenplate, J.T. and Fishhoff, D.A. (1990). Insect resistant cotton plants. *Bio-Technology*. 8: 939–943.

Sreenivasan, S. (2001). Indian Cotton: Implication of TO current quality scenario and measures for improvement *National J. Pl. Improv*., 3: 1–5.

7

Bio Safety Issues Related to Transgenic Crops

The developments in genetic engineering of plants around 1983 promised that this new biotechnology would benefit and even revolutionize agriculture. The transfer of desirable genetic traits across species barriers has the potential for solving problems in the management of agricultural crops (Anonymous, 2000). Potential benefits include reduced toxic pesticide use, improved weed control resulting in less tillage and soil erosion, efficient water conservation and improving the nutritional value of food crops.

The methodology employed today for gene transfer is not precise. However scientists can control with relative exactness the "trait gene" that is inserted into a host plant genome, but they cannot yet able to control its location, and the number of copies that get inserted. Multiple copies may "silence" the genes, resulting into instability in the introduced genes. Because Scientists cannot control the location and copies of the genetic material which are inserting, they cannot entirely control the traits of the host plant which will express the genes. In this way they cannot give the guarantee for stability in subsequent generations (Ryan *et al.*, 2001). This can lead to unpredictable and undesirable effects, examples of which include plant infertility, the production of toxins, and reductions in yield and plant fitness.

Among the unintended effects as a result of such gene transfer is a phenomenon called *pleiotropy*. Pleiotropy means that one gene controls multiple traits in an organism. In other words, the gene for a desired trait may also control other traits in the crop plant. Genetic engineers have relied on a simpler model of gene expression, assuming one gene controls one particular trait. However pleiotropy

The uses of genes for antibiotic resistance as selectable markers have also raised concerns regarding the transfer of such genes to microorganisms and thereby aggravate the healthy problems due to antibiotic resistance in the disease causing organisms. Although the probability of such transfer is extremely rare, steps are being taken to reduce this risk by phasing out their use. For *Bt* gene the vertebrates have been found to be lacking receptor binding site for Cry protein, hence, their health hazards are out of Cry toxin purview and Cry proteins are not toxic to vertebrates.

Risk to Environment

Assessing the environmental impact of GM crop is often difficult as many factors are considered. Scientists focus on the potential risks of GM crops and also emphasize their potential benefits. Risks to environment due to release of transgenic crops include *i.e.* the introduced genes may outcross with weedy relatives and have potential to create weedy species, direct effect on non-target organisms, effect on biodiversity and development of insect resistance.

Ecological scientists have little doubt that *gene flow* from transgenic fields into conventional crops and related wild plants will occur. Gene flow from transgenic to conventional crops is of concern to farmers because of its potential to cause herbicide resistance in related conventional crops. For example, in western Canada, three different herbicide-resistant canola varieties have cross-pollinated to create canola plants that are resistant to all three types of herbicide (Ellstrand and Norman, 2001).

Gene flow from transgenic crops to wild relatives creates a potential for wild plants or weeds to acquire traits that improve their fitness, turning them into "super weeds." There is already evidence of such out crossing from herbicide-resistant wheat to jointed goat grass. Other traits that wild plants could acquire from transgenic plants that would increase their weediness are insect and virus resistance (Erwin *et al.*, 2001).

Research conducted before the commercial application of the gene, and experience during five years in USA show that Cry1Ac binds to a specific receptor [170 kDaAPN Binding site in *H. virescense* Luo *et al.*, 1997] in the midgut of the sensitive insects, but does not affect mammals or insects that do not have the receptors,. The presence of the specific toxin receptor binding sites makes the Cry1Ac specifically toxic to a particular group of insects affecting mainly the lepidopteran insects and particularly, the cotton bollworm *H. virescense.* All other insects, wild life, marine life are not affected. The vertebrates have been found to be lacking receptor binding site for Cry protein, hence, out crossing is unintentional breeding of a domestic crop with a related plant. A major environmental concern associated with GM crops is their potential to create new weed through out crossing with wild relatives, or simply by persisting in the wild themselves. The potential for the above to happen can and is assessed prior to introduction and is monitored after the crop is planted as well. Study initiated in 1990 demonstrated that there is no increased risk of invasiveness or persistence in wild habitats for GM crops for the traits tested and compared to their unmodified counter parts (Anonymous, 1994). A recent study (De Nijs *et al.*, 2004) shows that only very limited effects on the environment have been detected in relation to out crossing. It is therefore important; that regulations are required to evaluate individual GM crops both prior to release and after commercialization.

The major environmental consequence resulting from the massive use of *Bt* toxin in cotton or other crops occupying a large area of the agricultural landscape, is that neighboring farmers who grow crops other than cotton, but sharing similar pest complexes, may end up with resistant insect populations colonizing their fields. This is because the lepidopteron pests that develop resistance to *Bt* cotton, moves to adjacent fields where farmers use *Bt* as a microbial insecticide (Gould, 1994). So emphasis has to be laid on studying the impact of transgenic crops on birds, mammals and soil biota.

Effects on Non-Target Organisms

It is a known fact that due to application of broad- spectrum insecticides many non-target, beneficial insects are also eliminated. Hence, their population is greatly reduced. It is also evident that *Bt* cotton is toxic to only some lepidopteran insects and do not possess adverse effect on non-lepidopteran, beneficial fauna. In Hebei province of China, cotton growers observed nearly 25 per cent increase in natural enemies of pests in *Bt* cotton fields [Nakanishi, 1999]. There was increase in population of natural predators such as wasps, aphid consuming flies and Lady bugs [*Hippodamia* sp] over that in non-*Bt*, conventionally sprayed plots [Buranakanonda, 1999]. Similar studies in cotton have been reported by Benedict *et al.* (1996) and Lambert *et al.* (1997).

Sears *et al.* (1999) and Losey *et al.* (1999) reported that pollen from *Bacillus thuringiensis* (*Bt*) insect resistant corn has a negative impact on Monarch butterfly larvae. This report raised concerns and question about potential risks to Monarchs and perhaps other non-target organisms. However to reach any conclusion these initial results are not sufficient and need conformation. Hansen-Jesse and Obrycki (2000) also reported higher rate of mortality in the monarch butterfly larvae feeding on milkweed larvae's naturally dusted with pollen from *Bt* corn plants. The study carried out by Wright *et al.* (2000) reported that common type of *Bt* corn had no deleterious effect on black swallow tail butter flies. These reports indicate the need to have more in depth research in this area both in field and laboratory conditions.

By keeping pest population at extremely low levels, *Bt* crops can starve natural enemies as these beneficial insects need a small amount of prey to survive in the agro-ecosystem. Parasites would be most affected because they are dependent on live hosts for survival and development, where as some predators thrive on dead or dying prey. Studies conducted by Birch *et al.* (1997) suggested that aphids were capable of sequestering the toxin from *Bt* crops and transferring it to its Coccinellid predators, thus, in turn affecting reproduction and longevity of the beneficial beetles. The potential of *Bt* toxins moving through food chain poses serious implication of natural bio-control in agro ecosystem.

A report from the US Environmental Protection Agency (EPA) indicated that data provide a weight of evidence indicating no unreasonable adverse effects of *Bt* proteins expressed in plants to target wild life. Crawley *et al.* (2001) has also concluded that in most commercial hybrids *Bt* expression in pollen is low, and laboratory and field studies show no acute toxic effects at any pollen density that would be encountered in the field. Lab experiments confirms that transgenic crops with insect resistance may have negative impact on beneficial insects, predators including lacewings (Hilback *et al.*, 1998), lady bird beetles (Birch *et al.*, 1997), monarch butter fly larvae (Losey *et al.*, 1999) and soil biota (Watrud and Seidler, 1998).

Impact of Insect Resistant Crops

The transgenic crops engineered with *Bt* genes (derived from *Bacillus thuringiensis*) offer protection against lepidopteran pest complex and may reduce the usage of synthetic insecticides that are used for control of insect pests. Since most of the crops have a diversity of insect pests than lepidopterans, insecticides will still have to be applied to control these pests which are mot susceptible to the endotoxin proteins expressed by the *Bt* crops. Most *Bt* toxins have a similar mode of action and once resistance develops to one toxin it can confer resistance to other related toxins, which is known as cross- resistance (Gouldburg *et al.*, 1992). The management of resistance in transgenic crops is in the experimental stage. However, based on previous experience some conceptual strategies have been advocated for effective management of resistance in transgenic crops. For resistance management, the important strategies are protein pyramiding, planting refuge row and Cry gene crop rotation. The field level tactic is one, which appears to be practicable to the farmers and can be readily practiced while the other strategies require high scientific intervention.

The scientist recognize that insects will develop resistance to *Bt* toxin hence use of refuge was always considered an interim solution. Now multiple gene constructs with another insertion of Cry gene apart from Cry1Ac (Voth, 2001; Penn *et al.*, 2001) is available. The new Cry gene in Bollgard II is Cry2Ab. The genotype of Bollgard II has shown promise for improved bio-efficacy and increased spectrum of control (Voth *et al.*, 2001). Cry2Ab expression has been found to be more than ten times as compared to the level of Cry1Ac in Bollgard II plants. The high plant expression of Cry2Ab has contributed to higher efficacy against important lepidopteron insects in cotton. In China, Wang *et al.* (1998) have also transformed cotton using construct consisting of P-lec and cowpea trypsin inhibitor genes for the control of bollworms. Hence, in future cotton varieties with multiple gene constructs will be a common feature and Schular *et al.* (1998) have given wide range of genes of bacterial, plant and animal origin which are effective against lepidoptera and some against coleoptera.

Bt has been widely used as a microbial spray because it is toxic only to caterpillars. In fact, it is a pest management tool that organic farmers depend on–one of the few insecticides acceptable under organic rules. Unlike the commercial insecticide spray, the *Bt* engineered into crop plants is reproduced in all, or nearly all, the cells of each plant, not just applied on the plant surface for a temporary toxic effect. As a result, the possibility that transgenic *Bt* crops will accelerate insect pests development of resistance to *Bt* is a serious concern. Pest resistance to *Bt* would remove this valuable and environmentally benign tool from farmers' and forest managers' pest control toolbox.

Safety Assessment of Transgenic Crops in India

Safety assessment of a transgenic crop is the most important step in this development process. Extensive testing and a long approval process includes comprehensive analysis of the risks and their scientific management to ensure food, feed and environmental safety before introduction into the market.

Safety assessments of a transgenic crop start with determining whether the product is substantially equivalent (except for defined differences) to conventional varieties. Further analysis then focuses on the evaluation of the defined differences by assessing potential safety risks of the host plant, gene donor(s) and the protein introduced.

Experiments are designed to identify the hazards, to assess risks and to take steps to manage the risk by applying logical strategies. Information on the following aspects needs to be generated on a case–to–case basis.

1. Characteristics of the donor organisms providing the target gene such as identification, pathogenicity, toxicity and allergenicity, the geographical origin, distribution pattern and survival mechanisms and the method of transfer of its genetic material to other organisms.

2. Characteristics of the vectors used such as the origin, identity and habitat, sequence and frequency of mobilization.
3. Characteristics of the transgenic inserts such as the specific functions including the marker gene inserts the expression levels, and the toxicity of the expressed products on the host plant, humans or animals.
4. Characteristics of the transgenic plant including methods of detection of the transgenic traits in the environment, toxicity and pathogenicity of the transgenic plants and their seeds to other plant, human and animals, possibility of the extent of transgenic pollen escape and pollen transfer to wild relatives and the impact on the environment.
5. Toxicity and allergenicity data are generated using standard protocol devised by national and international agencies.

Important precautions may be taken for minimizing the risks at release of transgenic plants

1. Special separation for isolation to prevent reproduction/ fertilization and seed setting.
2. Biological prevention of flowering by making use of sterility properties.
3. Human intervention for reproductive structures of flowers.
4. Controlling and destroying volunteer plants from experim-ental field.
5. Controlling the reproductive structures of transgenic plants like seeds and the plant prop gules from unaccounted spread.
6. Appropriate training of field personnel while handling the transgenic plants.
7. To take into account the proximity to human activity incase the transgenic plants have allergenic properties to human and animals.

All the data generated by the developing organization is then submitted in detailed formats to the government for seeking permission for commercial release. The initial risk assessment in India begins at the institutional level itself. The Institutional Bio safety Committee evaluates the proposal following which it has passed on to Review Committee on Genetic Manipulation and then Genetic Engineering Approval Committee. Another round of assessment with respect to agronomic benefits is undertaken under the ICAR system. There is provision for continuous monitoring by Monitoring and Evaluation Committee even after release of the variety at the center and state levels.

Bio Safety Regulation in India

A national Bio safety regulatory system to regulate production and release of genetically modified organisms (including transgenic crops) is considered essential in any country with a biotechnology programme. Bio safety regulations cover assessment of risks and the policies and procedures adopted to ensure environmentally safe application of biotechnology.

The Ministry of Environment and Forests, Govt. of India notified the rules and procedures for the manufacture, import, use, research and release of GMO's as well as products made by the use of such organisms on December 5, 1989 under the Environmental Protection Act 1986 (EPA). The rules 1989 order compliance of the safeguard, through voluntary as well as regulatory approach and any violation and non-compliance including non-reporting of the activity in this area would attract punitive action provided under the EPA. The two main agencies identified for implementation of the rules are the Ministry of Environment and Forests and the Department of Biotechnology, Govt. of India. The rules have also defined six competent authorities and composition of such authorities for handling of various aspects of the rules.

The six competent authorities as per rules are:

- Recombinant DNA Advisory Committee (RDAC).
- Review Committee on Genetic Manipulation (RCGM).
- Genetic Engineering Approval Committee (GEAC).
- Institutional Bio safety Committee (IBSC).
- State Bio safety Coordination Committee (SBCC).
- District Level Committee. (DLC)

Out of these, the three agencies that are involved in approval of new transgenic crops are

1. IBSC set-up at each institution for monitoring institute level research in genetically modified organisms.

2. RCGM set-up at DBT to monitor ongoing research activities in GMO's and small-scale field trials.
3. GEAC in the Ministry of Environment and Forests set-up to authorize large-scale trials and environmental release of genetically modified organisms.

Recombinant DNA Guidelines 1990

With the advancement of research in biotechnology initiated by various Indian Institution and industry, Department of Biotechnology had formulated Recombinant DNA Guidelines in 1990. These guidelines were further revised in 1994 to cover R&D activities on GMO's, transgenic crops, large–scale production and deliberate release of GMO's plants, animals and products into environment, shipment and importation of GMO's for laboratory research. Appropriate safeguards in handling organisms, plants and animals in various risk groups have been recommended.

For containment facilities and bio safety practices recommendations in the WHO laboratory safety manual on genetic engineering techniques involving microorganisms of different risk groups have been incorporated there in.

In 1998 DBT brought out separate guidelines for carrying out research in transgenic plants called revised guidelines for research in transgenic plants. These also include the guidelines for toxicity and allergenicity of transgenics seeds, plants and plant parts. The guidelines include complete design of a contained green house suitable for conducting research with transgenic plants. Besides this it provides the basis for generating food safety information on transgenic plant ant plant parts.

To monitor the impact of transgenic plants on the environment over a period of time, a special Monitoring-cum-Evaluation Committee (MEC) has been set-up by the RCGM. The committee undertakes field visits at the experimental sites and suggests remedial measures. The committee also collects and reviews information on the comparative agronomic advantages of the transgenic plants and advices the RCGM on the risks and benefits from the use of transgenic plants under evaluation.

There is an urgent need to create public awareness regarding not only the benefits of transgenic crops but also on their perceived risks.

There is also an urgent need of an in-depth study to address the effects of transgenic plants on non–target animals, insects, plants and other organisms etc, since not much work being done in this area.

After commercialization of transgenic crops in India, their impacts in relation to environmental and biodiversity should be monitored on regular basis to comprehend the impact on biodiversity in long run.

A comprehensive survey of the crop species, existing wild and weedy relatives in an environment into which the transgenic crops are intended to be released, need to be conducted. The information regarding degree of crossability between crop species and their wild relatives will help in devising better risk assessment procedures and appropriate strategies to minimize the risks in order to preserve the quality of an environment.

References

Anonymous (2000). Experiment Station and Extension Committees on Organization and Policy (ESCOP and ECOP). 2000. Agricultural Biotechnology. Critical Issues and Recommended Responses from the Land-Grant Universities. p. 20.

Anonymous (2005). Bio safety issues related to transgenic crops. Prepared by Biotech Consortium India Ltd in Association with Ministry of Environment and Forests Govt. of India. pp. 1–89.

ACAC (2001). Second generation of *Bt* cotton is coming. *ICAC Recorder*. 19(1): 3–7.

Alteri, M.A. (2000). The ecological impacts of transgenic crops on agro ecosystem, health. *Ecosystem Health*. 6: 13–23.

Bendict, J.H., Sachs, E.S., Altman, D.W., Deaton, W.R. and Kohel, R.J. (1996). Field performance of cottons expressing transgenic Cry1A insecticidal proteins for resistance to *Heliothis virescens* and *Helicoverpa zea* (Lepidoptera Noctuidae). *J. Econ. Entomol.*, 89(1): 230–238.

Buranakanonda, A (1999). *Bt* cotton is potential economic saviour of Chinese provinces. Bankok Post 26 Oct. 1999. p. 6.

Ellstrand, Norman C. (2001). When transgenes wonder, should we worry? *Plant Physiology* 125: 1543–1545.

Ervin, David E., Batie, Sandra S., Chantal Line Carpentier, and Rick Welsh. (2001). Transgenic Crops and the Environment: The Economics of Precaution.

Invited paper, Western Agricultural Economics Association. July 10. pp. 1–12. Logan, Utah.

Gouldburg, R.J. (1992). Environmental concerns with the development of herbicide tolerant plants. *Weed Technology*. 6: 647–652.

Gould. E., Martinez, R.A., Anderson, A., Ferre, J. and Moar, W.J. (1992). Broad spectrum resistance to *Bacillus thuringiensis* toxins in *Heliothis virescence*. *Proc. Natl. Acad. Sci. USA*. 89: 7986–7990.

Gould, F. (1994). Potential and problems with high dose strategies for pesticidal engineered crops. *Biocontrol Science and Technology*. 4: 451–461.

Hansen, Jesse. and Obrycbi, J.J. (2000). Field deposition of *Bt* transgenic corn pollen. Lethal effects on monarch butterfly. Oncology 19 August.

Hilbeck, A.M., Baumgartner, P.M. and Bigler, F. (1998). Effect of transgenic *bacillus thuringiensis* corn-fed prey.

Lambert, A.L., Bradley, J.R. and Vanduyn, J.W. (1997). Interactions of *Helicoverpa zea* and *Bt* cotton in North Carolina. *In* Poc. Beltwide Cotton conference, New Osleans, LA 7–10 Jan. 1997. National Cotton Council, Am. Memphi T.N. pp. 870–873.

Khadi, B.M. (2003). Commercialization of *Bt* Cotton. Its success and problems in Indian agriculture. Silver Jublee Lecture. Indian Society for Cotton Improvement. pp. 1–36.

Losey, J.E., Raynor, L.S. and Carter, M.E. (1999). Transgenic pollen harms monarch larvae [scientific correspondence]. Nature. pp. 399–414.

Luo, K., Sangadaia, S., Masson, L., Mazza, A., Brousseau, R. and Adang, M.J. (1997). The *Heliothis virescens* 170 kDa aminopeptidase functions as "receptor A" by mediatging specific *Bacillus thuringiensis* CryA d-endotoxines binding and pure formatio. *Insect Biochem. Mol. Biol.*, 27: 735–743.

Penn, S.R., Barb, R., Janson, O., Kris, E. and Greenplate, J.E. (2001). Quantification of lepidopteran activity in a gene product. A two year summary of Bollgard II. Proceedings of the Beltwide Cotton Conference. National Cotton Council of America, Memphis USA.

Ryan, Angela and Mae-Wan, Ho. (2001). Europe's New Rules Could Sink All GMOs. A report from the Institute of Science in Society, London. Aug. p. 28.

Schuler, T.H., Poppy, G.M., Kerry, B.R. and Desholm, I. (1998). *Insect resistant transgenic plants. Tibtech*. 91: 168–173.

Voth, R.D., John, T., Greenplate, J.E. and Mullions, J.W. (2001). Bollgard II cotton technical review. Proceedings of the Beltwide Cotton Conference, National Cotton Council of America. Memphis USA.

Wang, W., Chen, W., Gao, Y. and Zhu, Z. (1998). Obtaining of insect resistant cotton by transformation with Agrobacterium. Proceedings of World Cotton Res. Conference-2, held at Athens, Greece. pp. 350–354.

Watrud, L.S. and Seidler, R.J. (1998). Non-target ecological effects of plant, microbial and chemical introductions to terrestrial systems. Soil Chemistry and Ecosystem Health. Special Publication. 52. Soil Science Society of America, Madison, Wisconsin. pp. 313–340.

8

Pests and Their Management

Cotton crop is highly prone to attack of various insects and provide substratum to about 1326 species of insects from sowing to maturity in cotton growing areas of the world. About 162 species attain the status of pest responsible for enormous loss to cotton yield. This chapter deals with the identification of important cotton insects and scope of Integrated Pest Management to control the menace of these insect attacks. There are three broad categories of insects in cotton *i.e.* sucking insects, bollworms and other insects.

Sucking pests

Cotton leaf hopper (Jassids)

Identification

***Egg*:** Elongate ,yellowish white eggs are laid by moths.

***Nymph*:** Nymph is wedge shaped, pale yellow/green in colour and possesses wing pads covering first five abdominal segments. It has the ability to move forward and sideways on the under surface of leaves.

***Adult*:** Adult is wedge shaped, green in colour and bears conspicuous black dot on vertex of forewings and is very active flier (Fig. 6a).

Damage symptoms

Both the nymphs and adults suck the sap from the leaves. While sucking the sap, these insects inject toxins in leaf tissues, responsible for yellowing and curling of leaves from margins (Fig. 6b). The attacked leaves later turn reddish brown to produce hopper burn. The jassid starts damaging the cotton crop, as humidity and temperature conditions are high. In North India, the damage is triggered with the onset of first shower of monsoon during the June end or first week of July.

White fly

Identification

***Egg*:** Egg is stalked, Elliptical and light yellow when laid, Later on changes to dark brown.

***Nymph*:** Nymph is oval, light yellow and bears bristles on the body. In the later instars crawler, all other nymphal instars are immobile and remain attached to the under surface of leaves (Fig. 8a).

***Pupa*:** Pupae are elliptical and dark yellow with dark brown opercula.

***Adult*:** The adults are 1–1.25 mm in length with white wings having poor wing venation. The body of adult is covered with white waxy powder secreted through the vasiform orifice present on the upper surface of the body (Fig. 8b).

***Damage symptoms*:** White scale like nymphs remain attached to the undersurface of leaves and continue to drain the cell sap. The yellow spots become visible on the upper surface of the cotton leaf (Fig. 5c). The adults also suck the sap from the lower surface of leaf. Once the plant is disturbed, a cloud of insects flatter out and rapidly settles down on it after making whirling movements. Both the adults and nymphs while sucking the sap secrete honeydew on the upper surface of leaves, on which black '*sooty mould*' develop. The attack of whitefly gives the plant a sickly appearance. In addition to lowering of vitality of cotton, whitefly also acts as vector of leaf curl virus. The pest is serious in dry weather during poor monsoon season.

Aphids

Identification

***Egg*:** Aphids breed parthenogenitically and eggs are not laid. Females generally lay nymphs.

***Nymph*:** Nymphs are small, ovate, soft bodied and generally green/yellow in colour.

***Adult*:** Soft bodied insects are usually 1–2 mm long, wingless, black green with cornicles and cauda blackish in colour. Eyes are black.

***Damage symptoms*:** Both the nymphs and adults suck the sap from the young laves and tender shoots in clusters. The affected leaves show curling and distortion of the leaf lamina (Figs. 7a, b). In addition, honeydew is secreted on the upper surface of leaves making substratum for the development of black fungus called '*sooty mould*'. Initially, the upper surface of the leaves is sticky to touch in the morning hours. Black 'sooty mould' appearance is more peculiar on upper surface of middle canopy leaves. The pest is destructive to crop during the fag end of the crop season and destroys the lint in open bolls with honeydew and black 'sooty mould'

Thrips

Identification

***Egg*:** Eggs are kidney shaped and inserted into the tissue.

***Nymph*:** Nymphs are wingless and generally smaller in size.

***Adult*:** Adults are yellowish brown, 1 mm in length and legs are bladder like. Males are wingless, while females posses fringes of setae on the front and hind margins of wings.

***Damage symptoms*:** Both the nymphs and adults first lacerate the tissues with rasping and sucking type of mouthparts and later feed on the oozing cell sap. The feeding of thrips causes bronzing of leaves, peculiar from the lower surface of leaves. The attacked plants give bushy appearance with lopsided growth during the seedling stage. With the quickening of cotton growth with first irrigation or with the onset of rains, the damage diminishes to a greater extent.

Red cotton bug

Identification

Egg: Eggs are oval in shape and pale yellow when freshly laid and later on turn yellow and finally change to orange colour.

Nymph: The nymphs are bright red bugs with black wing pads and are 10–13 mm long.

Adult: Wings of adult are reddish in colour with a black spot/bar near the middle. Adult is generally 12–15 mm long.

Damage symptoms: The adults and nymphs suck the sap from the developing seeds inside the boll. Generally, no outer symptom of damage is available but these bugs cause yellow staining of lint. Water soaked spots on the inner wall of boll. The staining of lint inside the boll is mainly caused by the fungus (*Nematocera* spp.) entered through stylet punctures of bugs. Red coloured bugs fall on the ground crawl on the plant, if disturbed.

Dusky cotton bug

Identification

Egg: Eggs are cigar shaped and whitish in colour when freshly laid and later on turn pale yellow.

Nymph: Newly hatched nymphs are 2.5 cm long with rostrum extending beyond abdomen. Near first moulting, the colour changes to orange and then become reddish brown and thereafter more dark after each moult.

Adult: Adults are dusky in colour and measure about 5–7 mm in length. Its anterior is spined.

Damage symptoms: Both adults and nymphs feed on ripe seed. The pest inserts strong tube like stylets through seed coat into embryo and sucks the sap, affects the weight and viability of seed. They may also stain the lint due to crushing during ginning.

Bollworm pests

Spotted bollworm

Identification

***Egg*:** Eggs are spherical in shape and bluish in colour

***Caterpillar*:** The larvae is stout spindle shaped bears two pairs of fleshy, finger like tubercles. Though colour is slightly variable, yet are, *Earias insulana* has cream colour, Body marked with orange dots on prothorax. *E. vittella* is brownish in colour possessing median longitudinal stripe but without finger shaped possesses (Figs. 2a, 2b).

***Pupa*:** They are enclosed in tough silken cocoon, which are dirty or light brown in colour (Fig. 2c).

***Adults*:** The moth is 20–25 mm across the wings .The adult moth of *E. insulana* possesses green/yellowish green fore wings, whereas the hind wing appear pale. On the moth of *E. vittella* has green fore wings marked with green streak running medially from base to the outer margin (Fig. 2d).

***Damage symptoms*:** The caterpillars bore in to the growing shoots, buds, flowers and bolls. As soon as the terminal shoot of young plant is bored, the growing tip loses its turgidity and drops down. The bunchy branches developed on a plant due to the death of terminal. The caterpillar causes excessive shedding of fruiting bodies. The damaged squares (flower buds) open out to produce flared squares. These flared squares can fall from the plant with slight disturbance. Unlike the holes of American boll worm larvae, circular holes produced by the spotted bollworms

In the fruiting bodies remain filled with excreta. Damage to the crop seen on both vegetative and reproductive phases.

American bollworm

Identification

Egg: Eggs are green white ribbed and dome shaped. The colour of eggs change to dark brown /black before hatching.

***Caterpillar*:** The stoutly built caterpillar of variable colour possesses prominent noticeable dark brown broken lines along the body. The full grown caterpillar is about 40 mm long (Figs. 1a, 1b).

***Pupa*:** Pupae are brown in colour.

***Adult*:** Adults are pale brown yellow nocturnal moths with wingspan of 40 mm. Forewing are provided with series of black dots. The undersurface of forewings carries a black kidney shaped dark. The hind wings are yellowish brown with broad black outer band (Fig. 1 d).

***Damage symptoms*:** The caterpillar damages cotton leaves buds (squares), flowers and bolls. Unlike spotted bollworms, clean circular holes bored into the fruiting bodies are free from excretory pallets/ frass. These excretory pallets can also be easily spotted out on the soil surface under the attacked plants. There can be more than one hole per boll. Like spotted bollworms, the caterpillars of the bollworm also produce flared squares (bracteoles of flower buds open out as a result of the attack), which ultimately become yellow, die and shed with slight jerk (Fig. 1c).

Pink bollworm

Identification

***Egg*:** Eggs are elongated and flattened in shape and pearly white in colour when freshly laid and become deep brown near hatching.

***Caterpillar*:** The dark head of the newly hatched white caterpillar is distinct. Full grown caterpillar is white, 10–12 mm long and possesses a double pink band on upper surface of each abdominal segment. The stationery caterpillar appears apparently pinkish in colour from upper surface (Figs. 3a, b).

***Pupa*:** It is light brown in colour and later on turns brown.

***Adult*:** The adults are deep brown moths with wing span of 15–20 mm. Head and thorax are lighter in colour, forewings bear black spots, while hind-wings are silvery grey furnished with fringes of long hairs.

Damage symptoms

The caterpillar stage is destructive to buds, flowers and boils. The infested flowers can be identified from their typical, rosette appearance, harboring pink bollworm larva smeared with pollen. Sometimes, the premature boll/bud shedding can be seen on the soil. The larvae damage the seed. If the boll is cut opened, pink caterpillar can be seen in the damaged lint. The secondary rotting of attacked boll responsible for staining of seed cotton is an common feature of its damage. Late in the season, pink larva in diapauses stage in double seeds can be spotted out. The pest is more destructive to crop during the reproductive stage (Figs. 3c, d).

Other Insects

Cotton leaf roller

Identification

***Egg*:** Eggs are elongate in shape and are pale in colour

***Caterpillar*:** Agile caterpillar measures about 20–30 mm in length and is translucent green with black head and provided with thorasic shield.

***Pupa*:** Radish brown in colour

***Adult*:** Adult measures about 28–40 mm long across the spread of wings. The yellowish moth possesses black and brown spots on the head and thorax. Dark brown wavy lines on wings are the typical identification mark of moth.

Damage symptoms

The larvae generally feed on leaf tissues inside the rolled leaves starting with leaf margins. The infestation is particularly high under shady trees growing on bunds etc.

Cotton grey weevil

Identification

***Egg*:** Eggs are light yellow and round in shape

***Grub*:** Grubs are white cylindrical and legless.

***Adult*:** Adult beetles measures about 5 mm long and is grayish in colour. When adults are physically touched, they fall on the ground immediately and appear as dead.

Damage symptoms

Both the stages *i.e.* grub and adult are destructive in nature. Grubs fed on roots but the cotton plants are rarely killed. Adult weevils feed on the foliage and eat away the margins of the leaves and squares, Thereby making the outer boundaries of these parts irregular. The pest is present on the cotton crop throughout the season but is of minor importance.

Cotton Leaf Miner

Identification

***Egg*:** Eggs are oval elongate close to one another into the leaf tissues.

***Larvae*:** The full grown larvae is greenish in colour and measures 3 mm in length and possess smoky-brown mouth parts. It is pointed towards the mouth end and bluntly rounded from other end.

***Pupa*:** Brown pupa is in the larval mine for 7 to 15 days.

***Adults*:** Its tiny fly with grayish black mesonotum and yellowish frons with a pair of transparent shining wings.

Damage symptoms

The larvae feed on epidermis by making prominent whitish zig zag tunnels in leaves. Usually, the mine starts from the periphery and terminates towards mid-rib and contains one larva. The attacked leaves wither and dry up. The tunnels interfere with photo-synthesis of plant. Damage is confined young cotton crop. Generally lower leaves get affected as a result of the attack.

Cotton stem weevil

Identification

***Egg*:** It is globular, oval or cylindrical, milky white in colour and about 1.25 cm in length

Grub: It is creamy white in colour with slightly curved body with distinct head

Pupa: Pupa is white becomes brown before emergence of adult weevil

Adult: Weevils are dark brown in colour having black and whitish markings and about 3 mm length

Damage symptoms

The grubs attack the cotton plant during seedling stage. The attacked plants dry up immediately. However, in the slightly grown up plants, the-grub enter into the stem just at the ground level and start feeding by making tunnels inside the stem and result in swollen stem at the point of attack. The constriction of stem just above the soil is conspicuous at the soil level. Dried up plants can be easily spotted out in the field. The plants fall down with slight jerk or strong wind. The grubs can be easily identified after cutting the attacked stem.

Tobacco caterpillar

Identification

Egg: Eggs are dome shaped and are laid in clusters which are covered with hairs.

Caterpillar: They are light brown first and later on turn dark (Figs. 4a, b).

Pupa: Pupa is brown and about 10–12 mm in length.

Adult: The moths are 20–40 mm long with wingspan of 30 mm. The forewings are brown with small round rust coloured spots in the middle and beyond there are smaller kidney shaped spots.

Damage symptoms

The caterpillars are very gregarious and first feed near the place of hatching and later on migrate to the other leaves. They damage leaves, flowers and bolls.

Cutworm

Identification

***Egg*:** Eggs are dome shaped and have 30–40 longitudinal lines.

***Larva*:** The green caterpillars can be identified from their greasy but smooth skin. They have pale stripes and numerous black tubercles. The full-grown larva is dirty green or blackish and about 35–45 mm in length.

***Pupa*:** They are found in soil and are of uniform brown colour.

***Adult*:** The straw coloured moths are provided with brown kidney shaped spot on the forewings in apical region.

Damage symptoms

Caterpillars are destructive to young cotton crop. The larvae cut the young seedings at the ground level and the cut portion of the stem with young leaves can be spotted out lying near the left over stem of cotton on the soil surface. The removal of soil around the damaged plant exposes greasy caterpillars. (Please see Table 17).

Natural Enemies (Predators)

Lady bird beetle

Identification

***Egg*:** Egg is elongated and yellowish in colour.

***Grub*:** Body of grub is tapering posteriorly, black and marked with orange spots (Fig. 9a).

***Pupa*:** It is light brown marked with black dot and fixed at the posterior end.

***Adult*:** Adult is half pea shaped, yellow to reddish brown with 7 black spots on elytra, both the head and ventral side are blackish in colour (Fig. 9b).

Green lacewing

Identification

Egg: Eggs are pale green, void in shape with long pedicel which darkens before hatching.

Larva: They are alligator shaped and mottled grey or yellowish grey. When full grown they attain 2.0–2.5 cm in length.

Pupa: Larva pupates in protected place and spin globular parchment like cocoon and transforms into pupa.

Adult: Adults are medium size 1–2 cm long green yellow or sometimes brownish with golden eyes and delicate, netted wings.

Preying mantis

Identification

Egg: Ootheca is large, rounded or oval, yellowish white, spongy, attached to a twig etc. contains a large number of egg clusters.

Nymph: Small, pale green, wingless, resembles the adult.

Adult: About 2.25 inches long, greenish, elytra with a white patch tipped with brown, forelegs raptorial.

Spider

Identification

A body consists of two distinct regions, Prosoma and opisthosoma, the two being connected by a narrow waist or pedicel. The prosoma is covered by a uniform carapace bearing not more than eight simple eyes. The head is marked off by a groove. The opisthosoma is usually unsegmented, soft, rounded and without a telson, bearing not more than 4, usually 3 pairs of spinnerets. The chelicerae are two jointed, subchelate and contain the poison-glands. The pedipalps are six jointed, simple, leg-like, and form the copulatory organs in the male. Important species are trap-door spiders (*Pachlomerus*), funnelweb spiders (*Agelena*), wolf spiders (*Lycosa*), orb-web spiders (*Argiope*), jumping spiders (*Saltiicus*), crabspiders (*Misumena*), black widow spider (*Latrodectus*), etc.

Factors responsible for attraction of insect pests on cotton

- Cultivation of susceptible and uncommented with different habits of growth.
- Long duration of crop providing continuous substratum for these pests to thrive.
- Favorable green and succulent foliage.
- Favorable hot and humid climate for development of different pests.
- Nectaries on leaves and fruiting bodies as food substrates for insects.
- Availability of large number of floral forms (squares, buds, flowers and bolls)
- Indeterminate nature of cotton plant.
- Intensive cultivation.
- Non judicious use of irrigation and fertilizers.
- Indiscriminate use of insecticides without taking into account economic thresholds.
- Inadequate spray technology to produce thorough coverage.
- Non availability of really effective bio control organisms/ products.
- Extremely poor adoption of non insecticidal methods of pest management.
- Economic status of cotton growers does not match the recommended technology.
- Sole dependence of farmers on traders for pesticides which hinders the proper transfer of technology.
- Development of resistance in insects to insecticides due to indiscriminate use.
- Non availability of quality inputs to farmers also accentuated the pest.
- Excessive use of synthetic pyrethroids helped in buildup of sucking insect pests.

INTEGRATED PEST MANAGEMENT

To manage insect pest and diseases, it is always desirable to follow integrated pest and disease management to harvest economic yields of cotton. Following tactics of integrated pest management would be useful.

Table 1: Abiotic factors suitable for build up insect pests and period of activity of pests.

Insect pest	*Ideal weather*	*Period of activity*
Jassid	Hot and humid weather	July–November (Aug.–Oct. & Dec.–Jan.)
Aphid	Hot dry climate during vegetative stage and low temperature during boll maturation	July–November (Aug.–1st half-Sept.)
White fly	High temperature coupled with drought condition during late vegetative phase	Sep.–Nov.
Thrip	Hot and dry weather	Aug.–Dec. (Sep.)
Spider mite	Hot and dry weather	Aug.–Sep.
Cotton semilooper	Hot and humid weather	Aug.–Oct.
Cotton leaf roller	Hot and humid weather	Aug.–Oct.
Spotted bollworm	Worm and humid weather	July–Dec.
American bollworm	Worm and humid weather	Aug.–Oct.
Pink bollworm	Hot and humid weather	Oct.–Jan. (Nov.–Dec.)

Months in parentheses indicate peak activity

Cultural Control

Adoption of recommended agronomic practices during the crop season as well as in off-season at appropriate time helps to reduce the incidence of pest and disease. The cultural practices not only help in management of disorders but also create niches for conservation of natural enemy complex.

- *Planting time:* Proper selection of planting time enables the crop to escape damage by altering dates to the disadvantages of pest/disease.
- Reduce food availability to pest.
- Bring about synchronization between susceptible stage of the crop and damaging stage of pest.

Table 2: Status of cotton pests

Common name/ Scientific name	***Order family***	***Crop stage***	***Status***
Leaf minor/ *Leriomyza trifoli* Burgess	Diptera Agromyzidae	Cotyledionary stage	Minor
Cotton grey weevil / *Myllocerus undecim-pustulatus maculosus* Desbrocher	Coleoptera Curculionidae	Vegetative	Minor
Cotton leaf hopper/ *Amrasca biguttula biguttula* Ishida	Hemiptera Cicadellidae	Vegetative	Major
Aphid /*Aphis gossypii* Glover	Hemiptera Aphidae	Vegetative	Major
Thrip / *Thrips tabaci* Lindemann	Thysenoptera Thripidae	Vegetative	Minor
Whitefly/*Bemisia tabaci* (Gennadius)	Hemiptera Aleyrodidae	Vegetative	Minor
Spotted bollworm / *Earias insulana* (Boisduval) *Earias vitiella* (Fabricius)	Lepidoptera Noctuidae	Vegetative & Reproductive	Major
American bollworm/ *Helicoverpa armigera* (Hubner)	Lepidoptera Noctuidae	Vegetative & Reproductive	Major
Pink bollworm/ *Pectinophora gossypiella* (Saunders)	Lepidoptera Gelechiidae	Reproductive	Major
Red cotton bug/ *Dysdercus cingulatus* Fabricius	Hemiptera Pyrrhocoridae	Later stage	Minor
Dusky cotton bug/ *Oxycarenus laetus* Kirby	Hemiptera Lygaeidae	Vegetative	Minor
Semilooper/*Tarache notabilis* (Walker) *Anomis flava* Fabricius	Lepidoptera Noctuidae	Vegetative	Minor
Tobacco caterpillar/ *Spodoptera litura* Fabricius	Lepidoptera Noctuidae	Vegetative	Minor
Cottin leaf roller/ *Sylepta derogata* Fabricius	Lepidoptera Pyraustidae	Vegetative	Minor

Table 2: *(Contd...)*

Table 2: *(Contd...)*

Common name/ Scientific name	***Order family***	***Crop stage***	***Status***
Spider mite / *Tetranychus telarius* Linn.	Acarina Tetranychiidae	Vegetative	Minor
Semilooper/*Plusia acuta* Walker	Lepidoptera Noctuidae	Vegetative	Stray
Semilooper /*Plusia orichalcia* Fabricius	Lepidoptera Noctuidae	Vegetative	Stray
Termite / Odontotermes obesus (Rambur) *Microtermis obesi* Holmgren	Isoptera Termitidae	Vegetative	Minor
Pempherulus affinis Faust.	Coleoptera Curculionidae	Vegetative & Reproductive	Stray

Source: Anonymous (2005)

Table 3: Natural enemies of cotton pests

Common name	***Scientific name***	***Order–Family***	***Crop stage***	***Status***
Ledy bird	*Coccinella septempunctata* Linn. *Menochilus sexamaculatus* (Fab.)	Coleoptera–Coccinellideae	Early July-Sptember	Potential predator
Greenlace wing	*Chrysoprela carnea* Stephens	Neuroptera–Chrysopidae	Mid July–November	Potential predator
Syrphid fly	*Syrphus serarius* (Wiedmann)	Diptera–Syrphidae	August–Early October	Potential predator
Dragonfly	*Platygomphus dolobratus* Selys	Odonata–Gomphidae	June–December	Potential predator
Pentatomid bug	*Cantheconidea furcellata* Wolff	Hemiptera–Pentatomidae	Late July to November	Minor, Larval parasitoid
Trichogramma	*Trichogramma chilonis*	Hymenoptera–Trichogramma tidae	August –late November	Minor, Eggs parasitoid
Ichneumonid Wasp	*Ecphorosis perditincutus*	Hymenoptera–Ichneumonidae	August–November	Minor, larval parasite
Braconids	*Bracon gelechiae* Ashmead	Hymenoptera–Braconidae	August–November	Minor, larval parasite
Apanteles	*Apanteles flavipes* (Cameron) *Apanteles colomani*	Hymenoptera–Braconidae	August–November	Minor, Larval parasite

Source: Anonymous (2005)

- Reduces the number of generation of pests/disease cycles
- Planting time to disadvantages of pests and diseases
- Recommended seed rate to cover up the loss due to pests and diseases
- Recommended varieties tolerant/resistant to different insect pests/diseases
- Seed treatment against pests and diseases
- Proper plant to plant and row to row spacing
- Judicious use of fertilizers and pesticides
- Adequate irrigations for timely termination and high productivity
- Trap crops to divert pressure of pests from cotton
- Intercropping to mitigate incidence of different pests and diseases
- Destruction of alternate weed hosts of insects and pathogens
- Cropping pattern of minimize damage of pests.
- Termination of crop to avoid infestation of late season crop
- Destruction of sources of carryover of pests and pathogens

The staggered sowing extends the crop growing period and serves as substratum for the buildup of pests and disease continuously for longer period. As a consequence pests are able to complete 1–2 extra generations in the season. Alternatively, early planting of cotton can escape the severe damage of American bollworm. Pink bollworm incidence was comparatively more in early and late sown cottons Myrothecium incidence was also more in late sown crop in this zone.

***Plant spacing*:** The plant spacing has tremendous impact on the buildup of pests and diseases. The normal spacing provides optimum space for proper plant growth. The close spacing results bushy growth of crop that affects the penetration of light and also hinders the effective spraying. This well established fact that incidence of sucking pests and bollworms is comparatively more in closer planted cotton as compared to wider one.

***Fertilizers*:** The excessive use of nitrogenous fertilizers produces rank growth that increases the severity of sucking pests and bollworms. The optimum level of potassium and phosphorus is also essential for better yields. Phosphorus application to cotton limits the infestation of cotton whitefly.

***Irrigation*:** The adequate irrigation water is essential for optimum growth and creates natural resistance. The excessive uses of irrigation coupled with high nitrogen, close spacing and bushy type cotton genotypes results in unwanted plant growth that favor pest/ disease build up and also hampers pesticide efficacy. The judicious use of water is essential both during vegetative and reproductive phases of cotton to provide inbuilt resistance to insect pests/diseases. Late irrigation usually promotes the vegetative growth that is conductive for buildup of whitefly and American bollworm.

***Trap crops*:** The trap crop, which is the preferred host of pest for feeding, shelter and breeding, is exploited to attract and kill the pests species. Okra is recommended as trap crop to lure cotton jassid, spotted bollworms and American bollworm. Marigold grown as trap crops is recommended to collect and destroy the population of *H. armigera*. Planting of castor as trap crop is recommended to divert the population of *Spodoptera litura* from cotton.

***Cropping pattern*:** Cotton is mostly grown in rotation with wheat that is not a host of major insect pest of cotton, the inclusion of tomato in the cropping system, in recent years in cotton belt is also responsible for increased carryover of jassid, whitefly and American bollworm from one season to another. The other crop rotations, *i.e.*, cotton-groundnut, cotton-cowpeas or cotton-soybean are also less conducive for pest development on cotton. To manage cotton whitefly, less preferred hosts like maize, bajra and soybean is included in cropping sequence.

***Mixed cropping*:** Mixed cropping (cowpea + cotton) is being encouraged in M.P. to colonize natural enemies of cotton pests.

Destruction of Alternate Weed Host Plants

Several weeds are preferred alternate hosts of American bollworm, spotted bollworm and whitefly. The removal and destruction of these hosts in the vicinity of cotton fields, on sides of roads, etc., can help to reduce the multiplication and carryover of these pests.

Manipulation of Carryover Sources

The carryover of pink bollworm is through the un-pickable bolls/burns, cotton seed, shed material and gin trash, respectively.

To reduce the carryover following strategy is recommended to the farmers:

i. Allow the grazing of cotton field after the last picking.
ii. Removal of leftover bolls/crop residue.
iii. The larval mortality in cotton stacks put in the open and in upright position was higher. Therefore, the farmers are advised to stack the cotton sticks in the open away from tree shade in the villages in upright position.

Varietal Resistance

There have been constant efforts to develop varieties, which have resistance to major insect pests. Host plant resistance is now rated as top priority for integrated pest and disease management. It can prove highly effective based on risk cost -benefit analysis. The changes in pest scenario have taken place during the last decade. With the result, whitefly and American bollworm attained the status of key pests. Very effective chemicals failed to manage these pests resulting in failure of cotton crop. The change of pest scenario has created new challenges in evolving pest/disease resistant varieties. The development of varieties resistant/tolerant to a number of pests or a group of pests cannot be achieved easily.

***Biological control*:** Of the large number of natural enemies screened against insect pests/diseases, a few species were found economically important in regulating pest populations. Bio control agents are considered as vital in IPM and can be successful where pesticide use is minimal. Against insects, potential species of egg, larval, egg-larval, and pupal parasitoids have been recorded from cotton bollworms. *Trichogramma chilonis* an egg parasitoid is being used on large scale as component of IPM against cotton bollworm (*Helicoverpa armigera*) in south zone of the country. In an undisturbed cotton ecosystem, cotton pests are generally kept under check and will require insecticide interventions only late in the season. Many general predators like coccinellids, green lacewing, predatory bugs and spiders have also been found associated with cotton crop. Of these only *Chrysoperla carnea* can be exploited commercially for the management of cotton pests. In undisturbed ecosystem natural enemies keep the pests under check. In this context, the conservation and augmentation of natural enemies can play a significant role in sustaining the cotton productivity. Microbial pesticides containing nuclear polyhdrosis virus (NPV) and bacteria, *Bacillus thuringiensis* Berliner (*Bt*) have been utilized in control of cotton pests but with minimum success throughout the cotton growing areas. The use of NPV against H.*armigera* gave good success as components of IPM. However, the availability of area specified virulent strain, standardized and stable formulation, and technology for proper use under field condition are major constraints in adopting it as component of IPM. For management of diseases *Trichoderma harzianum* showed promise against pathogens inflicting root in cotton.

Chemical control: The pesticides will continue to be a major component of IPM owing to several constraints in use of cultural practices, availability of resistant cultivars and efficient parasitoids and microbes. In the absence of above factors, pesticides quantum would further increase and accentuate the problems of pests and diseases in cotton. The indiscriminate use has already created a number of new problems and enhanced the cost of plant protection and narrowed down the economic returns. According to available figures cotton consumes to the extent of 55 per cent of pesticides used in India.

EMERGENCE OF PESTICIDES RELATED PROBLEMS

1. Selection of wrong pesticides and their dosages.
2. Un-timed use against pests and disease.
3. Faulty application system.
4. Spurious pesticides in the market.
5. Resistance status in insects and pathogens.
6. Ignoring the negative role of pesticides in cotton ecosystem in time and space.

Bio-efficacy of pesticides

Sucking pests

Dimethoate and oxy-demeton methyl have been in use extensively against sucking pests since the last many decades and their efficacy has decreased in recent years and more effective nitroguanidine compounds like imidacloprid, thiomethoxam and acetamiprid are likely to replace such less effective molecules. Triazophos, ethion, phosalone, acephate and monocrotophos are identified as promising insecticides for management of whitefly. The indiscriminate and non judicious use of monocrotophos and acephate can lead to resurgence of whitefly.

Bollworm complex

Organochlorinated, (DDT, BHC and endosulfan) and carbamate (carbaryl) and organophopahtic insecticides (monocrotophos, phosalone, phenthoate and quinalphos) were in use for the control of pink and spotted bollworm up to 1980's. Synthetic pyrethroids (SP) were introduced in 1981 onwards and used extensively for bollworm management which revolutionized the cotton production. Subsequently, many new insecticides were recommended for the management of cotton pests.

Selection of insecticides

The success of insecticides is based on the selection of insecticides relevant to the pest complex of cotton ecosystem during vegetative (Jassid and whitefly) and flowering phase (pink, spotted American bollworms and whitefly). To manage the pests following guidelines must be adhered to:

1. Synthetic pyrethroids (SP) effectively manage the pink and spotted bollworms.
2. For *H.armigera*, conventional insecticides should be preferred. Chlorpyriphos or acephate should be used against grown up caterpillars.
3. Alternate use of synthetic pyrethroid and conventional insecticides is better for sustaining cotton production.
4. No insecticides should be used more than once during the crop season.
5. Restrict the use of SP if white fly is serious and that of quinolphos or chlorpyriphos, if jassid is serious.
6. In case of serious, multiple pest attack, the preference should be given to insecticides potent for *H. armigera*.

Economic Threshold Level

For judicious use of pesticides follow economic threshold levels (ETL) given against different insect pests of cotton (Table 4). To decide the right insecticides, dosages and time of application, pest monitoring is mandatory at macro level. Due to wide variations in insect population from fields to fields, emphasis is also given to the micro level surveillance. The field scouting should be done at random as per standard technique.

Table 4: Economic threshold level for insects

Insect	*Economic threshold level*
Jassid	More than jassid/leaf (minimum 20 leaves)
Aphid	10 aphids/plant or 15–20% plants infested
Thrips	10–15 thrips/leave or 15–20% plants infested
Whitefly	8–10 adults/leaf or 20 nymphs/leaf(minimum 30 leaves)
Spotted bollworm	10% incidence on fruiting bodies
American bollworm	10% incidence on fruiting bodies or 20 eggs/larvae or 20 plants
Pink bollworm	10% incidence on fruiting bodies

Source: Anonymous (1989); Anonymous (2005)

Spray Technology

Farmers must adopt efficient spray technology to provide adequate coverage to target sites. The use of high discharge nozzles, low and higher spray material, higher swath width and poorly maintained spray equipment adversely affected the bio-efficacy of insecticides which should be taken care off.

Botanical Insecticdes

Botanical insecticides azadirachtin enrich formulations (1%) or Neem Seed Karnel Extract (NSKE) 5% or Neem oil @ 4 ml per litre are effective against some cotton pests and must be substituted once or twice during early flowering stage in IPM.

Sex Pheromone

Sex pheromone of bollworms *viz. Earias* spp, *H. armigera*, *P. gossypiella* and *S. litura* proved extremely useful for monitoring, survey and surveillance, monitoring the population of bollworms, through sex pheromone can help in early detection of population and their build up.

Resurgence of Sucking Pests

The over reliance on insecticides for bollworm control has caused the resurgence of sucking pests. The indiscriminate use of synthetic pyrathroid resulted in white fly resurgence in cotton belt of the county. The severe out break of white fly were experienced in AP, Tamil Nadu, Gujarat and Punjab as a result of repeated use of synthetic pyrethroids. Besides this, the resurgence of aphid has also been noticed. The resurgence of mealy bug as result of the indiscriminate use of SP has been recorded in Tamil Nadu and Gujarat. Similarly the resurgence of mite was observed in Gujarat and Andhra Pradesh. (Please see Table 5).

Insecticide Resistance Management

The development of resistance in American bollworm and white fly to different insecticides is evident. The high degree of resistance to synthetic pyrethroid has been demonstrated from Andhra Pradesh, Tamil Nadu, Gujarat, Madhya Pradesh and Punjab. The Insecticides Resistance Management (IRM) strategy is really complex owing to the variation in crop cultivation and management strategies, cropping pattern from farmer to farmer in a region. Following majors are recommended for the judicious use of insecticides and to enhance the mortality of resistance strains of insect pests.

- Avoid insecticides intervention in cotton ecosystem early in the season for functioning of natural control.
- Cultivation of variety/hybrid resistance to avoid the use of insecticides in vegetative phase.
- Botanical or bio control agents should be encouraged as far as possible for management of insect pest in early phase of cotton crop (60–70 days after sowing).
- Judicious use of insecticides based on economic threshold particularly during flowering phase.
- Use of pest specific insecticides to reduces the resurgence of new insect pest

Table 5: Insecticides for the control of insect pests of cotton

Pesticide	***Dose (g ai/ha)***	***Current status***
	Bollworm complex	
Synthetic pyrethroids		
Cypermethrin	50–75	Effective against pink and spotted bollworms
Alphacypermetrhin	25–30	
Deltamethrin	10–15	Developed resistance to *H. armigera*
Fenvalerate	50–75	Indiscriminate use lead to white fly resurgence
Carbamate		
Carbaryl	1000–1250	Moderate control of bollworms
Organophosphates		
Acephate	1000–1500	Very effective against *H. armigera*
Cholrpyriphos	500–1000	Excessive use of acephate and cholrpyriphos
Quinolphos	500–750	Effective against *H. armigera,* Excessive use may be lead to the resurgence jassid
Profenofos	625–1000	Effective against *H. armigera*
Phenthoate	500–750	Moderate control of bollworm
Trizophos	500–750	Moderate control of bollworm
Ethion	1000	Effective against white fly
New Molecules		
Indoxacarb	75–100	Effective against *H. armigera*
Spinosad	75–100	Do
Emamectin benzoate	8–11	Do
Betacyfluthrin	18–75	Very effective against pink bollworm and spotted bollworm
Lambdacyhalothrin	25–30	Do
	Sucking pests	
Dimethoate	125–188	Low to moderate control of sucking pests
Oxydemethoate	125–225	Do
Monocrotophos	100–500	Do
Triziphos (White fly)	300–600	Very effective against whitefly
Ethion (White fly)	1000	Do
Acephate	1000	Excessive use cause resurgence
Imidacloprid	20	Effective against jassid
Acetamiprid	15	Do
Thiomethoxam	25	Do

Source: Anonymous (1989); Anonymous (2003)

- Alternate and restricted use of synthetic pyrathroids with other conventional insecticide and should be targeted for control of pink and spotted bollworms.
- Use of HaNPV for management of early instars of *Helicoverpa armigera* during early flowering phase.

- Restricting the use of insecticides to which the pest has developed resistance
- Avoid the use organophoshphatic insecticides till 90 days after sowing to enable the build up of natural enemies complex.
- Avoid the use of insecticides more than once in the season as far as possible.

Avoid the use of following cultural practices to reduce the incidence of insect pests on cotton:

- Delayed sowing especially of long duration varieties.
- Mono cropping of cotton to reduce the carryover of pests.
- Excessive use of nitrogen fertilizers and irrigation to check the bushy growth of cotton plants which favors pest build up and also hinders the proper protection of crop.
- Intercropping of okra, moong and pigeonpea.
- Deep ploughing during summer to expose the soil inhibiting stages of pests
- Destroy alternate weed host plant of cotton pests.
- Management of the pest on alternate crop like chillies, sunflower, tomatoes, tobacco, etc. should be done with non-insecticidal methods or with soft insecticides.
- Hand picking of larvae should be done wherever possible and larvae so collected can also be used for mass multiplication of NPV by the farmers themselves with simple available technique.
- Cultivate only recommended varieties with judicious use of fertilizer and irrigation water.
- Prefer to cultivate varieties with same growth habit in a particular area.
- Installation of bird perchers (8–10 per ha).
- Use of recommended insecticides at the recommended dosages.
- Avoid the use of insecticides, which cause resurgence or to which pest have developed resistance.
- Use proper spray technology with need based application
- Spray alternately synthetic pyrethroids and conventional insecticides as in case of pink bollworm and spotted bollworm

- Sometimes extensive dark green, water soaked lesions along the veins known as vein blight.
- Symptoms are usually more prevalent on lower leaves than on upper leaves.
- Lesions dry and darken with age and leaves may be shed prematurely resulting in extensive defoliation.
- Black lesions on the stem which girdle and spread along the stem or branch known as black arm.
- Pathogen inoculum may either be present in the field on infected crop residues from a previous season or it may be introduced at planting within infected seed. Lesions on cotyledons may be initiated by inoculum within the seed during germination.
- Inoculum from infected crop residues may be splashed onto the foliage and into the growing point of young seedlings.
- When environmental conditions are favourable the bacteria enter the plant via the stomata or wounds.
- Symptoms of bacterial blight develop when the temperature is over 25°C and relative humidity exceeds 85%. As lesions develop bacteria exude out onto the leaf surface for further dispersal through wind driven rain. The pathogen is able to enter the seed when mature, open, blight-infected bolls are exposed to wet weather prior to harvest.

Grey mildew (Ramularia areola)

- The disease generally appears on older leaves as the plants reach maturity, in the form of irregularly angular, pale translucent spots, 1–10 mm (usually 3–4 mm) in diameter and with a definite and irregular margin formed by the veins of the leaf (called areolate).
- The lesions are light to yellowish green on the upper surface. As the spots grow older, the leaf tissues turn yellowish brown while a whitish frosty growth appears chiefly on the under surface but occasionally also on the upper surface. This is the conidial stage of the causal fungus.
- Lesions occur on the bracts subtending the bolls. As the leaf becomes chlorotic, the lesion turns reddish brown and defoliation takes place.

- Early and severe defoliation leads to premature boll opening and immature lint. Minute, reddish brown flecks surrounded by a chlorotic band denote a hypersensitive reaction indicative of a high degree of resistance.

Leaf curl virus disease (ClCuv–Gemini virus)

- The initiation of disease is characterized by small vein thickening (svt) type symptoms on young upper leaves of plants. The disease is further characterized by upward curling of leaves which occur because of the uneven growth of veinal tissues on the abaxial side of the leaves. Later, formation of cup shaped or leaf laminar out growth called enations appear on the underside of the leaf. In severe cases and in plants affected at early age, reduction of inter-nodal length leading to stunting and reduced flowering/fruiting is observed.

Myrothecium leaf spot

The cotton crop in Madhya Pradesh is being severely affected by *Myrothecium* leaf blight during the last decade (Anonymous, 2003). The disease was first reported from Delhi in 1960 but it has been of minor importance during the last 30–35 years. The disease is however, gaining importance due to increasing incidence in almost all the states of the Central India. Although, the disease has been prevalent in Madhya Pradesh for long, systematic attempts have not been made earlier to quantify the prevalence and understand the intricacies of the disease.

The manifestations of the disease in Madhya Pradesh have not yet been documented. It has been observed that the symptoms are expressed on all aerial parts of the plant. Initially minute dark spots of 1–2 mm in diameter are formed which increase in size characterized by the presence of broad violet to brown margins surrounded by zones of translucent areas giving an appearance of concentric rings. Later, fructifications of the pathogen are formed as dark green coloured sporodochia with white mycelial margins arranged in concentric rings Adjacent spots sometimes coalesce giving blight like appearance. In some cases, the area covered by the necrotic lesions dried and withered giving rise to shot hole symptoms. On the bracts, typical violet coloured spots develop which gradually enlarges. On the green bolls, light brown coloured spots are formed which later turn into dark colour.

The disease has been occurring in severe form in the central region of India with a damaging effect on both the production and quality of cotton. There is an alarming situation of *Myrothecium* leaf blight in Madhya Pradesh both in terms of incidence and intensity as well. The disease has assumed a status of a major disease of cotton in Madhya Pradesh. During 2002–03 and 2003–04, the incidence of *Myrothecium* leaf blight was 41.71% and 38.28% respectively and the disease was observed in significant proportions in all the cotton growing districts of Madhya Pradesh. The pathogen has been identified as *Myrothecium roridum* with characteristic morphology. The symptoms have been documented. There was no variation in the symptoms under different agroclimatic situations and variability in the pathogen was not observed. All the stages of plants are susceptible. However, the plants seem to be highly susceptible up to 60 days. The conidial washing at 3000 and 6000 rpm before inoculation increased infection. There is significant increase in infection as the duration of wetness is increased. Injury by pin pricks was found to predispose the tissue to infection by *M. roridum*. *Myrothecium roridum* is seed borne in cotton and the preferred site in the seed was found to be the seed coat. Seed treatment with carbendazim (0.1%) was the most effective fungicide in controlling seed borne *M. roridum* and increasing the seedling vigour.

Maximum mycelial growth and sporulation was recorded at 28°C. and 88.5% relative humidity and a photoperiod of 24 hrs light/0 hr darkness. Maximum germination of spores was at 28°C and 100% R.H. with continuous light. The survivability of the fungus in leaf debris declined gradually in the first three months and declined sharply with a recovery of only 15% after 7 months. The transmission of the pathogen from plant to seed was 20.12% while from seed to plant was in the range of six to forty percent.

Sporodochia has been observed to be restricted up to the epidermis only. Aggregates of fungal mycelium forming stromatic masses in the intercellular spaces were seen. A hymenium layer is formed on the surface. Typical fructifications (sporodochia) are formed that are discoid/irregular shaped without setae on which the conidia are formed.

The incidence of the disease was higher in high rainfall situation irrespective of the soil type. The temperature, relative humidity and cumulative rainfall have a decisive influence on the expression of *Myrothecium* leaf blight and that the favourable climate is a mean air temperature between 24.5 and 28.8°C, relative humidity between 45 and 69% and cumulative rainfall in the range of 250.5 mm and 522.0 mm. The incidence and intensity of *Myrothecium* leaf blight influenced the seed cotton yield and the incidence exerts higher influence on seed cotton yield as compared to the intensity.

Four *G. arboreum* two *G. herbaceum* and two *G. hirsutum* germplasm lines have been identified as tolerant lines for use in future resistance breeding program. Carbendazim was found to be the best fungicide in reducing the radial growth, sporulation and germination of conidia. In field, all the five fungicides under study were significantly effective in controlling the intensity of *Myrothecium* leaf blight over control. The least percent disease index (PDI) was recorded for Carbendazim. The highest yield was recorded for Carbendazim treatment (894.51 kg/ha), which was significantly superior over all other treatments recording 25.42% increase in seed cotton yield over control.

Strategies that can be adopted for the management of the diseases

- Exclusion of the pathogen from the area – 'quarantine'. The pathogen from entering particular area where the disease is not prevalent.
- Elimination of alternate hosts/weed hosts. The pathogens of *verticillium* and *fusarium* wilts, *alternaria* leaf spot, bacterial blight and leaf curl virus have many weed/alternate hosts.
- Crop rotation with non-host crops.
- Crop residue management to eliminate the pathogens being carried over.

- Provision of balanced nutrition (*e.g.*) Potassium deficiency results in increased susceptibility to *alternaria* leaf spot and application of potassium increases the natural resistance of the host.
- Application of biocides through seed treatment and/or foliar sprays for control of *alternaria* leaf spot, grey mildew etc.
- Control of insect vectors–diseases caused by virus (*clcuv*) can often be controlled by controlling the vector (white fly–*Bemisia tabaci*) that carries the pathogen

Chemical control

- Carbendazim and the recently introduced triazole compounds *viz.*, propiconazole, hexaconazole, cyproconazole and tebuconazole and prochloraz (imidazoles) and benzothiodiazole (biological derivative) are effective fungicides in the management of grey mildew and *Alternaria* leaf spot and Myrothecium leaf spot diseases of cotton.

Biological control

- Biological control is an important area of focus.
- Biological control agents include – plant growth promoting rhizobacteria (*Bacillus*, *Burkholderia*, *Pseudomonas*, *Streptomyces*) and the fungal genera *Trichoderma*.
- There is growing awareness that biologically based pest management (BBPM) fitting in the existing IPM strategies provide more environmental friendly and economically viable alternatives for agriculture.
- Act by competitive exclusion, biochemical antagonism or induction of host defenses.
- Biopesticides are cheaper than synthetic toxins by 50 per cent.
- They are ecofriendly, have a high cost benefit ratio and do not induce resistance in plant pathogens.
- Several advantages of biological agents as seed treatment.
- The saprophytic nature of biocontrol agents makes large–scale production feasibility.

- Small amounts of inoculum requirement.
- Simple methods of application.
- Systemic spread along the surface of the developing root system.
- Antagonistic activity on the root surface during the economically important phase of early root infection by the pathogens.

Disadvantages of the Biopesticides

- Narrow spectrum of activity.
- Inconsistent performance in practical agriculture.
- Environmental sensitivity, and
- Short shelf life.

Direct Effects

- Biocontrol fungi *viz*, *Trichoderma* spp. suppress seedling disease pathogens on cotton in the field.
- Use of *Pseudomonas fluorescens* strains (pf1 and mmp) as seed treatment and foliar spray have been found to be effective in the control of bacterial blight (*Xanthomonas axonopodis* pv *malvacearum*) of cotton.
- The beneficial fungi like *Trichoderma harzianum* locate and attack the damping off pathogen *Rhizoctonia solani.*
- The beneficial fungal strands entangle the pathogen and release enzymes that dehydrate *Rhizoctonia* cells eventually killing them.

Indirect Effects

- Addition of compost with heavy cellulosic materials inhibits *Verticillium dahliae*, *Fusarium oxysporum* f.sp. *vasinfectum*, *Pythium* sp. In cotton due to better colonization and competitive saprophytic ability of antagonistic fungi.
- It has also been found out that the compost activates disease resistant gene which might be responsible for the reduced incidences.

Fig. 3: Pink bollworm - *Pectinophora gossypiella*

Fig. 3a

Fig. 3b

Larva stage

Fig. 3c

Fig. 3d

Damage symtomps

Fig. 4: Tobacco caterpillar - *Spondoptera litura*

Fig. 4a

Fig. 4b

Larva Stage

Fig. 4c Skeletomsed leaves

Fig. 4d Adult

Fig. 5: Cotton semilooper - *Anomis flava*

Fig. 5a: Larval stage

Fig. 5b: Adult

Fig. 6: Cotton Jassid - *Amrasea bigutlula biguttulu*

Fig. 6a

Fig. 6b

Fig. 7: Cotton Aphis - *Aphis gossypii*

Fig. 7a: Leaf Infestation

Fig. 7b: Damage symptoms

Fig. 8: Cotton Whitefly - *Bemisia tabaci*

Fig. 8a: Nymph

Fig. 8b: Adult

Fig. 8c: Leaf infestation

Fig. 9: Lady Bird Beetle

Fig. 9a: Larva

Fig. 9b: Adult

Fig. 10: Chrysopa sp.

Fig. 10a: Larva stage

Fig. 10b: Adult

Fig. 11: Syrphid Fly

Fig. 11a: Adult

9

Organic Cotton Cultivation: Future Prospects and Problems

Traditionally, Indian farmers have been following organic cultivation till the middle of the last century, as they had no other choice. The advent of high yielding varieties and hybrids of cotton during the last three decades has resulted in large scale dependency on chemical fertilizers and especially toxic pesticides. Cotton, cultivated on 5% of India's cultivable land consumes 54% of the total pesticides used in Indian agriculture (Mehta, 1996). This has not only increased cost of cultivation but also caused environmental pollution. Some of the chemicals marketed, particularly, for plant protection are highly dangerous to human health and also leave poisonous residues in soil after application lasting over a few years.

Global awareness of health and environmental issues is spreading fast in the recent years, especially in the developed countries. Sustainability in production has become the prime concern in agriculture development. Organic method of farming is the best option to ensure air, water and soil around us unpolluted, leaving the environment safe for the present and future generations.

A new trend is being developed in India now to produce some of the crops organically not only on account of the love for protection of nature but also because of the need for having safe products for human consumption.

Cotton enjoys an extremely positive image due to its naturalness and gentleness to human skin. Life cycle of cotton products from cultivation through production to disposal is embedded with harmful chemicals, which seems to impair human health and develop allergic illness. Realizing that millions of tonnes of chemicals are used to produce and process cotton and the likely effects these chemicals will have on human health and environment in the long run, it becomes imperative that organic cotton production is enhanced.

General Principles

Organic agriculture is a process for developing a viable and sustainable agro-ecosystem. The time between the start of organic management and certification of crops is known as the conversion period. When traditional agricultural methods fulfill the principles of the organic standards, no conversion period is required. When also claiming virgin land for organic agriculture, no conversion period is required. The whole farm, including livestock, should be converted according to the standards over a period of time. The certification programme defines this period as conversion period. In the case of field crops or annuals, normally the conversion period is two years and for perennial crops three years. If a farm is not converted all at once, it should be done on a field-by-field basis, whereby full standards are followed from the start of conversion on the relevant fields. The area of land being managed and meeting the certification standards will therefore progressively increase. It is not recommended to separate the conversion of individual fields unless this is on the basis of imposed public restrictions or can be justified according to local conditions. The compositions of a farm unit can vary widely according to geographical conditions, ownership structure, time span required etc. In the case of farms having both crop production and livestock, the entire operation should be seen as a whole.

Organic management prevents degradation of common biotic and abiotic resources, including areas used for rangeland, fisheries, forests, forage for bees as well as neighboring land, air and water. So organic agriculture follows viable and sustainable agro-ecosystem by working compatibly with natural living systems and cycles. Species and varieties selected and cultivated in organic agriculture should be adaptable to the local soil and climatic conditions and tolerant to pests and diseases. Biological and cultural means are followed to prevent unacceptable losses from pests and diseases. Crops and varieties should be suitable to the environment and there should be a balanced fertility programme to maintain required nutrient levels in the soil with high biological activity, locally adapted rotations, companion planting, green manure production and other recognized organic practices.

Organic cultivation improves structure and fertility of the soil through balanced choice of crops and implementation of diversified cropping system. Biological processes are strengthened without re-coursing to chemical remedies, such as, synthetic fertilizers and chemical pesticides. In this system control of pests, diseases and weeds are primarily preventive, and if required, by adopting organic products, which will not adversely affect the environment. Organic matter of various kinds, nitrogen fixing plants, bio–fertilizers, pests and disease resistant varieties, bio-control agents, soil improvement practices, such as, mulching, fallowing, crop rotation, multiple cropping, mixed farming etc. are freely adopted. In brief, organic farming merges traditional and respectable views on nature with modern insights.

Organic livestock husbandry is based on harmonious relationship among land, plants and livestock respect for the physiological and behavioral needs of livestock and feeding of good-quality organically grown foodstuffs. Organic management practices promote and maintain health and well being of the animals through balanced organic nutrition, stress-free living conditions and breed selection for resistance to diseases, parasites and infections. The conversion period from the conventional livestock husbandry depends upon the period to develop natural behavior, immunity and desirable metabolic functions. Organic animals are those born and raised on organic holdings. Breeds should be adapted to the local conditions. Such animals should receive their nutritional needs from organic forage and feed of good quality. Their transport and slaughter should be with minimum stress.

Organic aquaculture promotes biodiversity, biological cycles, biological activity and sustainability. Use of organic feed, adoption of proactive health management and provision of desirable living conditions are part of it. Standards generally do not allow the use of synthetic chemicals for biocides, colourants, antibiotics and growth hormones.

Genetic engineering is not allowed in organic production and processing. Accordingly Genetically Modified Organisms (GMOs) and their derivatives are excluded from organic production, processing and handling.

Organic Standards

The most significant factors distinguishing organic farming from other methods of sustainable agriculture are the existence of production and processing standards and certification procedures. The standards are developed by private associations, companies and certification bodies or by the State itself. Universally accepted standards have been prepared by the IFOAM. A number of regional, national and international standards have been developed worldwide.

Several countries have formulated and adopted or are formulating rules and regulations on organic farming, processing and certification requirements considering the national situations. The European Union developed their standards as early as 1991 and enacted under EEC Regulation No. 2092/91. The standards of the European Union are improved periodically by amending the regulation. The United States published their standards under the name, United States Department of Agriculture National Organic Programme (USDA NOP) and made effective from 21st October 2002 under the Organic Food Production Act of 1990, as a part of the 1990. Farm Bill. Japan introduced their organic standards under the title, Japanese Agricultural Standards (JAS) as per the Notification No. 59 of the Ministry of Agriculture, Forestry and Fisheries Certain countries like Switzerland, Argentina have also made their regulations for organic production and processing.

In India the organic standards came into existence when the Department of Commerce, Ministry of Commerce and Industries, Govt. of India notified them in March 2000 considering the export potential of organic agricultural products. These standards are put under the National Programme for Organic Production (NPOP). A logo has also been adopted for the organic products certified under NPOP called Indian Organic Logo. The National Steering Committee has been constituted with representatives from the Ministries of Agriculture, Food Processing Industries, Forest and Environment, Science and Technology, Rural Development and Commerce and Industries to act as the apex advisory body for assisting the Govt. of India in shaping the growth and development of organic production. The National Steering Committee will meet as often as necessary as but not less than twice a year. An accreditation committee has also been set up with membership to Tea Board, Coffee Board, Spices

Board, APEDA and other agencies as the executive arm of the NPOP for given accreditation to inspection and certification agencies.

Future Prospects

The concept of organic farming is based on holistic approach where nature is perceived to be more than just an individual element. In this farming system there is dynamic interaction among soil, humus, plant, animal, eco–system and environment. Hence organic farming differs from industrial agriculture as in the latter, biological systems are replaced by technical production systems using chemicals liberally.

India has a great potential for organic cotton. Nearly 70% of the Indian cotton area (6.4 m ha) is managed by resource poor farmers and they grow cotton under risky rainfed condition and farmers who cultivate cotton in general avoid application of costly chemical inputs like pesticides and fertilizers (Venugopal *et al.*, 1996). Out of total rain-fed area nearly 45% is under desi cotton which do not requires either knowingly or unknowingly follow the organic method of cotton cultivation. If these potential organic farmers are enlightened with the technology of organic cotton cultivation and premium price is ensured. Thus these traditionally cultivated desi cotton farmers can be channeled into a organic farming system.

The 68th plenary Meeting of the International Cotton Advisory Committee (ICAC) suggested that organic cotton has the potential to provide new ideas that can influence and support wider sustainability derives in the sector. The organic cotton sector continues to grow, albeit more slowly in the global economic slowdown, reaching over 180,000 metric tonnes of lint in 2008–09. Organic cotton is one of the cotton to meet the challenges of land use, food security and water scarcity. Small growers exposed to variable rainfall and problems from debt produce most organic cotton in India. Many farmers view organic cotton as a risk reduction tool, but the use of fewer purchased inputs involved a trade off with productivity. Often the smallest and most resource poor farmers come to organic, and in fact organic works better in marginal or tribal areas. An essential element of organic cotton production is to have more planning and certainly through partnership to ensure long term viability. Organic cotton production, processing, spinning and marketing at the retail level involve a long and complex chain requiring traceability and communication links between producers and consumers. The future organic cotton production may involve contract farming with direct links to retailers. Breeding cotton genotypes for organic farming is the new challenge to scientist (Narayanan, 2010).

In Madhya Pradesh despite cotton cultivation being highly dependent on intensive chemical uses, there are certain pockets like part of Khandwa, Jhabua and Burhanpur (tribal area) where cotton is grown more or less organically, that is without chemical inputs. Thus there is an enormous opportunity of organic cotton cultivation over a vast area of around 1 lakh hectares in general and in pockets like tribal belt of Nimar and Malwa regions in particular, where the traditional agriculture has not been exposed to modern cultivation practices. In these area farmers are not using pesticides to raise the crop. Though organic agriculture is traditionally linked with low yield, the productivity levels of these areas are encouraging under rainfed condition. This is despite the area having the problem of water scarcity. The farmers of these area, changing varieties every year, it is because of non availability of seed of suitable variety for these areas. The important recommended varieties for these areas are Jawahar Tapti and sarvottam in *G. arboreum* category and Khandwa-2 and JK-4 in *G. hirsutum* category. Despite several factors in favour of organic cultuvation in the country, the organic cotton production is still negligible in the national pool. This is mainly because of:

1. Lack of awareness among the farmers for organic cultivation.
2. Conventional farmers have psychological make up that organic production is risky. This is despite un-intentionally their growing cotton nearly organically.
3. Organic cotton production is subject to strict test of regulations in order to be certified by the certifying agencies.
4. Farmers have no exposure for the requisite procedures and norms adopted in organic agriculture.
5. Organic market itself is in fluid state with consumers having little knowledge on organic values while making purchase decisions.
6. Lack of clear cut workchart, on what criteria and above all also will render a green/organic certificate. This uncertainty is the major tripping stone.

Certification Problems

Organic produced cotton cannot be sold as "green" without certification. The present certification mode is very subjective, based on various inspections of the production material and methods of the registered farmers. Obviously, these standards vary from one agency to the other and need to be standardized. At present organic cotton cultivation proceeds according to the standards of International Federation of Organic agricultural Movement (IFOAM). In this context Govt of M.P. also involved in the promotion of organic cotton production and can also play a pivotal role in making the standards more applicable under local condition. They can also be vested with responsibilities to certify the produce. However, clear–cut standards have to be worked out for this purpose.

Suggestions for increasing organic cotton production

To tap the vast potential for organic cotton production, concerted research and development efforts are needed. The *G. arboreum* varieties under cultivation need to popularized among the farmers like Jawahar Tapti and JK–9 (Madhya Pradesh), AKA 5, PA 205, PA 402 (Maharashtra) and DLSA 17 (Karnataka) having high yield potential with inbuilt resistance mechanism against sucking pests. Still there is need to identify ideal plant type suitable for organic agriculture couple with improvement in biotic tolerance and yield sustaining management practices. Though little input like fertilizer are applied, yet one should remain prepared for low yield under such system. Strong will is desired as today consumers may not be prepared to pay extra for eco-friendly products but in future all corporate decisions are bound to be based on environmental values, both producers and consumers are needed to be educated on the subject. State Goverment has good setup of agriculture department and they can expect to play a pivotal role in this endeavor. Clear-cut measures as to who will certify on what criteria's, how much would be the demand and of course, who would be the buyer at what premium are some of the bottlenecks and need to be developed. Once that is finalized, it will be possible to increase production of organic cotton in a well–organized manner.

An innovative effort to grow organic cotton is also under way in Madhya Pradesh. Since 1992, more than 1100 farmers from 77 villages (Anonymous, 2001) in Khargone and Badwani districts of M.P. have been involved in a project called Maikaal Bio-Re-produce organic cotton. The area of cultivation extended to about 3265 hectares and the total production is estimated at about 3226 metric tonnes (Shastry and Koutu, 2002). The project is in collaboration with Maikaal Fibres Limited (India) and Remai A.G., Switzerland. The state government of Madhya Pradesh also promotes the organic cotton cultivation by providing technical know how.

Problems in Availability of Quality Organic Inputs

Quality control of organic manures

The awareness of the organic farming has increased with the consumer more concerned with the quality of the produce. Organic products also fetch a better price because of, this there is a great demand of organic manures. A wide variety of bio-degradable wastes are used in the preparation of composts, resulting in wide variations in nutrient content and quality of the final product. There is at present no methodology to monitor the quality of compost and no standards have been developed in respect of quality control at different level. Thus a number of spurious or sub standards products are being marketed, which is a major limitation on the efficient use of compost in organic agriculture.

Maturity of the compost is an important component of quality. Immature compost has adverse effects on crop like cotton. The toxic elements derived from the raw material can be harmful to plants. This is especially true in the case of raw materials from municipal wastes. They can have high level of heavy metals like As, Cd, Pb etc. Phytotoxic compounds such as volatile acids are produced when wastes are stored under anaerobic condition.

Regulation for quality control

This is important in commercial production to regulate the quality produce. Government control over commercial production is required.

Monitoring of quality control

The guidelines for quality control have no meaning unless there is strict monitoring. For the monitoring to be effective, some Govt. body has to be involved in periodic sampling analysis from production sites and marketing point in huge quantity. Some accredited laboratories have also to be identified in the analysis of organic manures for identified parameters.

Commercial organic manure bags should also carry clear and informative labels. Good labeled bags enable the consumers to compare different brands in terms of nutrients and other quality parameters. This would eliminate substandard products from the market.

Yield loss in organic cotton

Yield loss in organic cotton is inevitable. No comparative statistics is available to evaluate the loss. However ICAC has given the yield loss estimate in organic cotton cultivation in few countries.

Technology for organic cotton cultivation

Considering the low productivity and the intrinsic risk and uncertainties in the cotton production certain low input sustainable production technologies was evolved to augment and stabilize the productivity of cotton in the region. The component of organic production system includes farmyard manure/vermi compost, green manures, bio-fertilizers, bio-pesticides and bio-control agents for pest management.

Table 1: Organic cotton production per hectare (1993)

S. No.	*Country*	*Total organic production (tonnes)*	*Organic yield (kg/ha)*	*Convent-ional yield (kg/ha)*	*% Yield loss in organic production*
1	Argentina	1.8	290	451	–36
2	India	124.6	181	280	–65
3	Turkey	15.3	627	1009	–38
4	USA				
	a. Arizona	1338.6	1076	1366	–21
	b. California	3363.5	1076	1509	–29
	c. Tennessee	150.7	538	504	+7
	d. Texas	655.2	538	544	–1
	e. Virginia	1.1	544	709	–23

Source: ICAC 1994

Maintenance of soil fertility and nutrient management for organic cotton production

Improvement and maintenance of the organic matter in the soil is an essential precondition to sustain reasonable levels of organic cotton production. This would also increase water holding capacity, reduce erosion, improve soil structure, besides enhancing the supply of nutrients, particularly, NP and S. Hence a good organic matter management programme is needed for fertility management under organic production system. Rainfed cotton crop in Central India removes around 5.8 kg N. 2.0 kg P and 6.6 kg K per 100 kg seed cotton produced (Pundarikakshudu, 1985). With many such alternate uses of FYM, in such a huge quantities will required to meet the crops nutrient requirement which is generally not available. Hence, combination of sources with different biological properties must be resorted. The following components were standardized for integrated nutrient management under organic cotton production.

1. *Farm Yard Manure*: FYM @ 5 tonnes/ha with 0.6–0.8% N., 0.2–0.25% P with C:N ratio of 26–28% spread uniformly and harrowed 15 days prior to sowing.
2. *Use of Dhaincha (Sesbania) as green manures*: It is grown as green manure crop between two rows of cotton and turned down at 25 days after sowing. Its fast decomposing leaves provides N during the crucial early boll development period while the stalk acts as temporary mulch.
3. *Vermicompost*: It offers good scope for recycling of farm wastes. This is to be applied @ 1–1.5 tonnes/ha. It contains several divers microflora and enzymes that add in good plant growth.
4. *Use of Bio-fertilizers*: Continuous use of chemical fertilizers has resulted in soil degradation, multiple nutrient deficiencies and air and water pollution. To sustain soil health and high crop productivity, judicious use of different sources of nutrients such as chemical, organic and biofertilizers is essential. In this context bio-fertilizers assume importance as a component for organic cotton cultivation. Bio-fertilizers are cheap and eco friendly and can supplement other sources of nutrients to an extent of 25–30 per cent.

Azospirillum

Azospirillum, a microaerophyllic bacterium, is known to enter into associative symbiosis with xylem vessels of plant roots. Besides its ability to fix elemental nitrogen. Azospirillum is also known to secrete growth promoting substances like GA and IAA, which enhance root proliferation and growth of crop plants. They are known to add 25–40 kg N/ha/year. Azospirillum are effective both for seed treatment and soil application @2 kg/ac. Once the practice gets stabilized, farmers would be advised to increase its usages and reduce the usages of chemical fertilizers.

Phosphobacteria

Phosphobacteria, particularly those belonging to genera *Bacillus* and *Pseudomonas* possess the ability to solubilise insoluble inorganic phosphorus and make it available to plants. It has been estimated that about 30 kg/ha/annum is solubilised by their application. Phosphobacteria is very much important component of organic cultivation in India and state Govts, is providing it on subsidized rate. Soil application @ 2 kg/ha along with Azospirillum (2 kg/ha) is recommended

5. *Use of drought and insect tolerant varieties*: Use of drought and insect tolerant traditional varieties helps in sustainable cultivation of organic cotton production. Among the cultivated species of cotton *G. arboreum* varieties namely J. Tapti and Sarvottam are also found suitable for organic farming.
6. *Insect management system*: Unfortunately, in the last four decades, much emphasis has been given to chemical control at the expense of other alternative methods of control. The emergence of insect resistance to insecticides becomes a real problem which makes more people to appreciate the need for other alternatives. Perceiving the urgent need to give increased attention on this aspect it is essential to develop separate package of practices for eco-friendly insect management.

a. Mono cropping of cotton is found to be conducive for pest development than mixed cropping *i.e.* cotton–soybean or cotton–groundnut. These systems also help to increase the abundance of predators and parasites in the cotton system.

b. The very late seeded crop harbored high population of leafhopper. Similarly higher dose of nitrogen and late sown crop invite the invasion of jassid attack.

c. Removal of weed plants that acts as alternate hosts for the pests is of almost importance.

d. Removal and destruction of crop residues, helps in reducing diapausing leaves of pink bollworms.

e. Botanical Neem seed kernal extract and neem oil having an antifeedant action was found to be very affective against heliothis and whitefly.

f. Management of insects through natural enemies in the eggs and larva parasites like Trichogramma sps helps in reducing spotted, pink and heliothis bollworms during early reproductive phase of the crop. Among the different predators, lady bird beetle and *Coccinella* sps. were found to regulate the population of bollworms in the cotton ecosystem to varied degree.

g. Predatory birds also controlled the field population of heliothis bollworms up to 84%.

h. Application of NPV @ 250 LE/ha was found to be very effective in reducing the population of heliothis.

i. The application of *Bacillus thuringiensis* was also proved to be effective against heliothis.

j. Management of diseases through Biofungicides Pathogens like Pythium, Fusarium, Rhizoctonia etc., cause diseases like damping off, root rot and wilt. Application of *Trichoderma* spp., a bio–control agent, has been found to be an effective alternative to manage root diseases. Seed treatment with trichoderma @ 50–100 g/kg of seed followed by soil application @ 2.5 kg with 100 kg of FYM, twice at 30 day interval has been recommended.

k. Traps Pheromone and light traps were recommended. These traps help in monitoring of adult population of pests like bollworm complex.

l. Topping is recommended not only to check the unnecessary vegetative growth but also to remove eggs laid by insects on top shoots thus aiding in their non-chemical control.

m. Removal of cotton stubble after harvest, without opting for a second (Ratoon) crop or prolong the crop growth, is recommended to break the cycle of pests.

n. Bird perches are recommended to attract birds which feed on caterpillars and helps in their natural control.

Experience about the Production of Organic Cotton Using Variety Sarvottam

Progressive Cotton farmer Shri Om Prakesh Jain, Village Borasar Burhanpur MP was invited to share his experience of organic cotton cultivation during the Workshop on Eco–friendly Cotton, Organized by Indian Society for Cotton Improvement, Mumbai on 28th October (1998). He narrated that "in the month of May, 1995–96 a programme was held in our village by the M.P. state Agriculture Dept. in which they expressed their own experience regarding organic cotton [Sarvottam] cultivation under the guidance of Agriculture Officers. After this I decided to grow this cotton. Four kg seed was received from Khandwa research center. SADO Burhanpur provided details regarding organic fertilizers, culture and neem oil. Firstly I prepared my one acre land with the help of tractor deeply. Two trolleys of "Nadep Khad" were also used in this land which was prepared specially by me. Before sowing, the seed was treated with Thiaram 3 g/acre and germination was 90–95 percent. Square spacing of 45 × 45 cm was adopted. 15–20 days after germination, I used the method of inter-cultivation in these plants. Besides this, 3 kg PSB + 1.5 kg Azotobacter culture was given according to the advice of agriculture dept. Inter cultivation and other agricultural system which was useful for my crop was done by me. Agriculture Dept. officials visited my field many times and gave me appropriate guidance. When crop was 60 to 90 days old, I used Neem oil 5 g/l for spray. After rainy season in the end of the month of October, the plants were irrigated by me [two times]. In this way, I received 6.50 quintal cotton per acre on the basis of organic cultivation. The market value of cotton was 11250/– while expenditure was Rs. 3250/– and the profit was Rs. 8000/– per acre."

References

Allen, Will (1994). Organic cotton production: a niche market or a must market_ presented at the 'Cotton Connection' conference for an ecologically and socio-economically sustainable cotton production. Pestizid Aktions Netwer & AK Cotton Connection. November, 25–26. Hemburg, Germany.

Anonymous (1991). Council Regulation (EEC) No. 209291 of 24.6.1991 on Organic Production of agricultural products and indications referring thereto on agricultural products and foodstuffs. In official *Journal of the European Communities L.* 198: 1–15.

Anonymous (1994). Proceedings of Seventh Australian Cotton Conference. Queensland, August. pp. 187–202.

Caldas, T. (1994). Agro-ecological assessment. Initial recommendations. Pesticides Trust, Project Sustainable agricultural and rural development: sustainable cotton production system. West Africa Senegal. Report No. 1. August. p. 21.

ICAC (1994). Production of Organic Cotton -II. In ICAC Recorder, Vol. XIII, No. 2 June. pp. 17–21.

IFOAM (1993). The first International IFOAM Conference on Organic Cotton. Cairo/ Egypt. Sept. 23–25. p. 38.

Kannavar, S.K. (1995). Organic Cotton need of the hour. Clothesline, August 1995. p. 7881.

Kannan, L. (1996). Organic Cotton: Niche market or Paradigm Shift. Special issue on Eco-friendly Cotton. *J. ISCI.* 21(2): 238–245.

Lovisolo, Ruth. (1994). National standards and Certification of Organic produce. Proc. of the Seventh Australian Cotton Conference, Queensland. p. 571.

Mor, Omprakash (1996). Experience on Cotton Production. Special issue on Eco-friendly Cotton. *J. ISCI.* 21(2): 147.

Murray, David (1994). Organic cotton on the Darling Downs: A pest manager's nightmare Proc. of the Seventh Australian Cotton Conference, Queensland. pp. 199–202.

Narayanan, S.S. (2010). Reorientation of cotton improvement models and strategies in practical perspectives for 21st century needs. *Cotton Research Journal.* (1): 1–17.

Noergaad, J. (1992). Green Cotton: a practical example of cleaner production-Ministerial Cleaner Production Seminar. Paris. October 27–29. p. 6.

Rajendran, T.P., and bambawale, O.M. (1994). IPM strategies for rain fed cotton cultivation in central India-presented at the 'cotton Connection' conference for an ecologically and socio-economically sustainable cotton production, Pestizid Aktions Netswerk and AK cotton Connection. November 25–26, Hamburg, Germany.

Rammamoorthy, K. (1995). Combating concerns about cottons Impact on the Environment – Clothes-line. pp. 98–99.

Regupathy, A. (1996). Pesticide residue in cotton ecosystem. Paper presented in National Workshop on Pesticide Application Technology for Cotton. CICR, Nagpur. pp. 19–20.

Venugopal, K. and Natarajan, K. (1992). Changing needs of IPM on cotton production in India. Paper presented at state level IPM Conference, TNAU, Coimbatore, Sep, 1–2, 1995. pp. 1–2.

10

Naturally Coloured Cotton

The cotton fibre which is the purest form of natural cellulose is bestowed with versatile properties for a wide range of ever changing textile fashions and needs. Cotton lint that is produced extensively in many countries of the world is mostly white in all the four cultivated species of *Gossypium*. Several colours in cotton are found to occur naturally in the genus *Gossypium* species *hirsutum* in addition to species of *arboreum*, *barbadense* and *herbaceum* since time immemorial (Sika and Joshi, 1960). The spectrum of colours in cotton comprise colours like brown, black, mahogany red, red, Khaki, pinkish brown, blue, dirty white, green and bright green (Singh *et al.*, 1993) and their various shades also.

The commercial cultivation of naturally coloured cottons, especially as "Eco–Friendly Cotton" has assumed special importance in some western countries to meet the challenges and health hazards increasing due to the effluent water of textile industries causing severe pollution of drinking water resources. It is needless to mention that the chemical dyeing process employed for the dyeing of fabric is discharging millions of gallons of highly toxic effluent water everyday containing very high concentration of heavy metal like Hg, Pb, As, Fe, Co etc., which is damaging the environment day by day. It is not merely a question of fashion of craze for fabric of naturally coloured cotton or of earning foreign exchange by creating attraction for coloured cotton, which is encouraging the research and development of coloured cotton around the world, but it is rather really a matter of concern for the biologist and agricultural scientist to realize the importance of naturally coloured cotton to protect the global environment and minimize/eliminate dangers of water and air pollution. However, the white cottons need due care and protection legislatively against any type of mechanical or genetical contamination. The research and development efforts on existing colour cottons are to be considered very important, of course not at the cost of white cottons, when the mankind has to choose the alternative whether to live and let live or continue the existing pollution creating modernized industrial progress without caring for the future generations.

It is in this context that exploitation of naturally coloured cotton assumes special significance as a sign of awareness for at least protecting drinking water for the mankind in future as well as for the life of animals, birds and aquatic creatures in the larger interest.

Genetics of Coloured Cotton

It has been reported that these colours have dominant type of inheritance over recessive white colour (Fletcher, 1907) and to be governed by multiple factors in addition to various modifiers and intensifier genes. The basic colour gene, a major gene, operates in association with modifiers that are suppressors or intensifiers located on closely linked loci. Often the genes for coloured lint proved allelomorphic to white and incompletely dominant to it (Christidis and Harrison, 1955). These genes have already been investigated, documented and mapped on the chromosome of the cotton genome long back by several research workers (Ramanathan and Balasubrahmanyan, 1933; Hutchinson, 1934, 1935; Silow, 1944). In *G. hirsutum* and *barbadense* two independent loci for lint colour have also been reported (Silow, 1944). The linkage groups in colour lint controlling genes in old and new World cottons and gene symbols assigned for various lint colours in all the four cultivated species and others have been reported by Endrizzi *et al.* (1985).

Genetic Resources of Coloured Cottons

Germplasm resources of coloured cottons are available in National Gene Banks of USA, India, China, Uzbekistan, etc.. In India, about 40–60 coloured cotton genotypes (genetic stocks) are reported to be available besides those of wild species and races (Narayanan and Sundaram, 1996). These represent both indigenous and exotic accessions from USA, Central Asia, Isreal, Peru, Mexico, Egypt, etc.

The coloured types in Indian gene resource include Algerian Brown, Arvin Brown, Texas Rust, LL.55–68–2, LC.1–1, Hirsutum Tashkant, Parbhani American, LC1–3, Brimer Browen, Lousiana Brown, Fuzzy Seed Brown, SA.65, LC.1, Red5-7, Brown Hirsutum, Cotton 162, Cotonnark Dark Brownand Light Brown, Kampala Red, Nankin Brown, Nankin Khaki, Extreem Okra, Arkansas Green, Arkansas Brown, Texas Brown, Intense Red GreenLint, KC. 94–2, KC–94–2–28 etc in *G. hirsutum*, Egyptian Brown, Green Lint, Pima Brown, etc in *G. barbadense* and Cocanadas 1 and 2, Khaki Arboreum, Red northerns, RH25, etc in *G. arboreum*. Genetic resources of coloured cotton available at National gene bank are presented in Table 1.

Table 1: Genetic resources of coloured cotton available at national gene bank

Sources	***No. of accessions with coloured lint***	***Lint colour***	***Exotic/ Indigenous***
Hirsutum	40	Brown, shades of brown, green	Both
Barbadense	2	Brown	Exotic
Arboreum	5	Brown, red-brown, khaki	Indigenous
Herbaceum	2	Brown	Indigenous
Wild Species,	17	Brown, brownish,	Exotic
Races	–	tan, grayish, creamy	

Source: Singh *et al.* (1996)

Based on colour intensity of lint, *Gossypium* species are grouped into different colour groups as follows:

Lint colour	***Species***
Brown	*G. aridum, G. hirsutum*
Browhish	*G. armourianum, G. mustelinum, G. sturtianum, G. stocksii, G. anomalum, G. hirsutum, G. australe, G. darwinii, G. somalense, G. capitis-viridis*
Brownigh green	*G. areysianum, G. gossypioides, G. robinsonii, G. herbaceum, G. harknesii, G. longicalyx, G. sturtianum* var. *nandeswarense.*
Tan	*G laxum, G. lobatum, G.trilobum, G. raimondii, G. hirsutum, G. incanum, G. arboreum.*
Tan creamy	*G. triphyllum*
Red Brown	*G. tomentosum*
Creamish	*G. barbadense*

Research and Achievements

Ms. Sally Fox (1987) is the first commercial cotton breeder to have developed American colour linted genotypes/cultures which are claimed to be comparable in all characters to white cotton. She has developed varieties with brown, green and red coloured lints and also multi-coloured lint such as green and white, off white and tan, and red and green. The commercial lines developed and patented for sale by Natural Cotton Colours Inc., are also becoming popular. However, no viable market has been established in USA for organic coloured cotton.

Reports indicate that olive green and fawn brown coloured cotton have been developed and are under cultivation in Israel. It was reported that blending them with virgin white cotton lint in suitable proportions can give various shades of processed yarn. The original colour was also maintained when washed. The yield of coloured lint was 700 kg.per hectare when grown in over 650 hectare as against 1800 to 1900 kg/ha. for white linted cotton in Israel. The coloured lint fetched 2 to 3 times higher price than white linted cotton. These Israeli coloured cottons possessed 26.3 mm fibre length with 3.8 to 4.5 micronaire value for geeen and 2.5 to 3.7 micronaire value for brown along with 19 to 21 g/tex fibre strength. Israeli cotton has been imported by a textile mill in India for pilot spinning and processing.

The brown colour from *G. richmondii* has been transferred to Texas Marker–1 in China and the resulting brown lint cotton was reported to have yield potential of 3000 kg/ha, but it has low ginning of 30%.

In India the work on development of naturally coloured cotton genotypes in *G. hirsutum* having high yield potential, resistance to pests and diseases with superior fibre and colour properties for spinning and fabric manufacturing purpose undertaken at JNKVV, Khandwa, has resulted in development of genotype KC–94–2 (JCC.1). This genotype was evaluated in large scale for commercial production in M.P (Mandloi *et al.*, 1996). The yarn of 30s count produced in small scale trials was found suitable for weaving into a shirting fabric with 66 ends/inch and 46 picks/inch. The fibre and yarn characteristics of this variety, first fibre test report of coloured cotton cultures KC-94

(CIRCOT), yield and yield component, fibre qualitry traits of *G. hirsutum* coloured cotton cultures developed by JNKVV and performance of coloured cotton culture KC–94–2 on Farmers' field trial conducted during 1995–96 are presented in Tables 2.1–2.5.

Table 2.1: Fibre and yarn characteristics of KC–94–2

Fibre characters	***Value***	***Yarn character at 30s nominal count***	***Value***
2.5% Span length (mm)	24.5	Nominal counts	30s
Uniformity ratio %	48	Corrected Lea CSP	1929
Micronaire value	3.7 to 4.6	Thin places (per 100 m)	43
Percentage mature fibres	77	Thick places (per 100 m)	145
Tenacity at nominal "O" gauge (g/tex)	36 to 38	Neps (per 100m)	122
		Single thread breaking strength (g)	207
		Tenacity (g/tex)	10.6
		Elongation (%)	5.9
		ASTM Appearance grade	D

Source: Mandloi *et al.*, 1996

Table 2.2: First Fibre Test Report of coloured cotton cultures KC–94 (CIRCOT)

S. No.	***Culture***	***Mean fibre length (mm)***	***2.5 % span length***	***Micronaire value***	***Tenacity at "0" gauge (g/tex)***
1	Mix sample (coloured cotton)	19.0	20.1	2.9	31.3
2	KC–94–1 (Shining pinkish brown)	22.4	24.4	4.1	36.4
3	KC–94–2 (Shining brown)	24.6	26.9	2.5	37.5
4	KC–94–3 (Shining brown)	20.1	21.6	2.0	33.1
5	Parent sample	22.1	24.1	3.9	34.0

Source: Mandloi *et al.*, 1996

Two cultures namely KC–94–10 and KC–94–12 have been identified at Khandwa to possess spinning capacity of 40s count. In addition to this, transfer of cytoplasmic male sterility to the coloured cottons is in progress.

Table 2.3: Yield and yield component traits of *G. hirsutum* coloured cotton cultures developed by JNKVV

S. No.	*Culture*	*Plant height (cm)*	*Duration (days)*	*Bolls/ plant*	*Boll wt. (g)*	*G.P.*	*Lint index (g)*	*Seed index (g)*	*Lint yield/ plant (g)*	*Seed cotton yield/ plant (g)*
1	KC-94-1-18	62	165	18	3.0	30.1	3.2	7.5	16	54
2	KC-94-1-20	60	160	12	3.2	30.5	3.8	7.6	19	62
3	KC-94-1-22	63	150	20	3.0	30.5	3.5	8.0	18	58
4	KC-94-1-24	50	155	28	3.2	30.5	3.5	7.9	24	80
5	KC-94-1-26	55	150	25	3.2	30.0	3.6	8.5	23	75
6	KC-94-2-18	65	160	18	3.5	35.5	4.0	8.0	19	56
7	KC-94-2-20	70	160	32	3.5	32.7	3.8	7.9	26	80
8	KC-94-2-22	82	160	31	3.6	33.8	4.5	8.9	25	75
9	KC-94-2-24	75	160	28	3.9	33.2	4.8	8.7	30	90
10	KC-94-2-26	78	160	30	3.8	32.8	4.1	8.5	30	90
11	KC-94-1-1	60	155	25	3.2	31.2	3.4	7.5	22	70
12	KC-94-2	62	165	28	3.2	33.0	3.7	7.5	26	80
13	KC-94-1-2	45	165	30	3.0	30.2	3.1	7.2	25	82
14	KC-94-1-1-8	75	190	38	3.2	33.5	4.4	8.8	34	102

Source: Mandloi *et al.,* 1996

In view of the growing awareness of environmental hazards and the ecofriendly nature of coloured cottons, research achievements in other countries and the sporadic demands for such cottons from a few foreign countries, investigations on coloured cottons have gained momention in India on arboretum cotton. UAS Dharwad also progress a lot on arboreum cotton with the development and release of

Table 2.4: Fibre quality traits of *G. hirsutum* coloured cotton cultures developed by JNKVV

S. No.	*Cultures*	*2.5 % span length (mm)*	*Fibre fineness (Micronaire value)*	*Maturity coefficient*	*Bundle strength (g/tex) at "0" gauge*	*Lint colour*
1	KC–94–1–18	22.4	3.2	0.64	35.9	Shining brown
2	KC–94–1–20	22.6	3.2	0.64	35.9	Shining brown
3	KC–94–1–22	23.4	3.3	0.65	34.8	Brown
4	KC-94-1–24	24.6	3.2	0.62	35.4	Light brown
5	KC–94–1–26	25.4	3.5	0.67	35.7	Shining brown
6	KC–94–2–18	22.1	3.0	0.67	37.8	Shining pinkish brown
7	KC–94–2–20	24.9	3.0	0.61	36.4	Shining brown
8	KC–94–2-22	23.9	3.3	0.64	39.1	Shining brown
9	KC–94–2–24	23.6	3.5	0.66	37.0	Brown
10	KC–94–2–26	25.1	3.2	0.62	39.1	Brown
11	KC–94–1–1	22.6	3.9	0.68	37.5	Light brown
12	KC–94 –2	21.3	3.5	0.67	34.8	Shining light brown
13	KC 94 1–2	23.4	3.2	0.64	37.0	Light brown
14	KC-94-1 -1-8	22.4	3.7	0.69	37.5	Shining brown
15	CAT–4020 –1 Coloured check	20.3	3.6	0.66	39.1	Dark brown
16	Khandwa–2 (white lint check)	26.0	3.7	0.71	43.4	Normal white

Source: Mandloi *et al.*, 1996

Table 2.5: Performance of coloured cotton culture KC–94–2 on Farmers' field trial conducted during 1995–96

S. No	*Name of farmers (sqm.)*	*Area taken*	*Plant popul-ation*	*Total prod-uction*	*Yield kg/ha*	*2.5% span length*	*Unifor -mity ratio*	*Streng th (g/tex) at"0" gauge*	*Finen-ess micr -onaire value)*
1	Beej Nigar Farm, Khargone	1000	900	90.5	905	–	–	–	–
2	Narendra K. Patel	2000	1400	223.5	1117	23.3	46.8	37.3	3.7
3	Rampal Patel	1000	600	177.0	1770	23.0	49.1	37.7	4.4
4	Shashikant Mehta	1000	1100	114.5	1140	24.4	46.7	37.0	3.2
5	Shirish Patel	1000	900	175.0	1750	24.7	47.0	37.1	4.0
6	Ekanth Choudhary	1000	1100	157.0	1570	24.3	49.0	37.5	4.2
7	Jayant Kishore Choudhary	1000	600	48.0	480	23.9	47.7	37.7	4.1
8	Kanhaiya-lal Patel	1000	600	78.0	780	24.3	47.3	37.2	3.9
9	S.K. Sharma	1000	700	45.0	450	23.9	23.9	37.7	4.1
10	Gajraj Malakar	1000	500	71.0	710	23.6	47.9	37.4	3.7
11	Richard Farm	1000	700	62.0	620	24.0	47.9	37.9	3.8
12	Poonamchand Kushwaha	1000	500	31.0	310	–	–	–	–
13	Swamy Vivekanand Ashram, Buldhana (MS)	1000	300	130.0	1300	–	–	–	–
14	Marl Overseas Ltd.	1000	500	123.0	1230	23.3	44.6	37.5	3.7

Source: Mandloi *et al.*, 1996

arboreum coloured cotton variety. Yield traits and fibre quality parameters of some of the colourd cotton accessions are presented in Tables 3.1–3.2

Table 3.1: Evaluation of coloured cotton germplasm cultures for yield and quality characters

S. No	*Germplasm lines*	*Seed cotton yield/ plant (g)*	*G.P.*	*Boll weight (g)*	*Halo length (mm)*	*Lint colour*	*Reaction to sucking pests*	*Source*
1	C–2	10	32.5	1.9	17.8	Green	S	CICR Combatore
2	C–9	06	33.0	2.0	16.0	Reddish brown	MS	–
3	C–6	08	33.1	2.2	15.5	Brown	S	–
4	C–8	06	33.5	2.4	21.1	Brown	S	–
5	CAT–1130	15	32.2	2.8	14.6	Brown	S	CICR Nagpur
6	CAT–868	16	33.6	2.2	21.4	Light brown	MR	–
7	CAT–4020–1	18	33.0	3.2	18.5	Dark brown	S	–
8	CAT–4020–2	20	33.1	1.5	20.9	Dark brown	S	–
9	CAT–1317	15	30.1	2.7	14.9	Brown	MS	–
10	CAT–1175–1	22	31.0	3.5	19.6	Light brown	MS	–
11	CAT–1135	18	32.8	2.8	21.9	Light brown	MS	–
12	CAT–3499 (SA–3)	10	32.2	2.8	20.6	Green	S	–
13	CAT–2334 (SA–5)	08	33.6	1.5	19.4	Dark brown	MS	–
14	H–202	07	31.4	2.6	22.9	Brown	MS	–
15	KC-94–2 (control)	22	33.0	3.7	20.9	Shining brown	MR	J.N.K.-V.V

S–Susceptible; MS–Moderately susceptible: MR–Moderately resistant
Source: Mandloi *et al.,* 1996

Important Varieties

Red Northerns

Red Northerns of Andhra Pradesh belonging to *G. arboreum* race *indicum* used to be grown on red soils and had dark red colour. Northerns cotton grown on black soils was a mixture of *G. herbaceum*

Table 3.2: Fibre properties of brown linted accession of *G. hirsutum* L. cotton

Accession	***Lint colour***	***2.5% span length (mm)***	***Uniformity ratio (5)***	***Maturity coefficient***	***Fineness (MV)***	***Strength (g /tex)***
Red 5–7	Dark brown	21.3	4646	0.80	3.6	4.7
Cotonark (DB)	Dark brown	24.1	46	0.62	2.7	40.7
Extreem Okra	Medium brown	15.1	49	0.83	3.9	
Kampala coloured light brown	Light brown	22.9	44	0.64	2.8	37.1
Nankin Brown	Light brown	20.0	51	0.76	3.4	34.6

Source: Singh *et al.* (1996)

and *G. arboreum* and was white in colour. In general, *indicum* cottons had good strength and a superior variety N-14, selected from white Northerns, eliminating *G. harbaceum* and Red Northerns plants, was released for cultivation in the tract.

Cocanadas 1

Cocandas 1 belonged to G.*arboreum* race *indicum*. It was released for cultivation during 1947 in coastal district of Andhra Pradesh. It had parrot green foliage and light brown lint colour. It was grown in Nellore, Guntur, Krishna, East and West Godavari districts, parts of Vishakhapatnam, Cumbum and Markapur, Talukas of Kurnool and Nalgonda districts of Andhra Pradesh.

Cocanadas 2

Cocandas 2 was evolved through pure line selection. It had brown colour lint. It was superior to Cocanadas 1 in intensity of lint colour and fibre quality. Cocanadas 2 was released for cultivation during 1951 for the Cocanadas 1 tract.

DDCC.1 (Dharwad Desi Colour Cotton1)

DDCC.1 is a *G. arboreum* pureline selection from local bulk population. It has been identified for growing in diploid white cotton tracts of Karnataka during 1999.

The economic characters and fibre properties of these four varieties are given in Table 3.

Table 3: Economic characters and fibre properties of diploid naturally coloured

Characters	*Red Northerns*	*Cocanadas 1*	*Cocanadas 2*	*DDCC .1*
Yield (kg/ha)	N.A.	662	682	800 to 1000
Ginning out turn (%)	22	28	30	30 to 32
2.5% span length (mm)	20 to 22	20 to 22	20 to 22	23
Micronaire value	N.A.	N.A.	5.2	5.0
Tensile strength at 3.2 mm gauge (g/tex)	N.A.	N.A.	N.A.	18 to 19
Spinning capacity (counts)	16s to 18s	18s to 22s	18s to 22s	
Lint colour	Reddish	Light brown	Brown	Almond

N.A: Data not available
Source: Mandloi *et al.* (1999)

Jawahar Coloured Cotton (JCC–1)

JCC1 is a *G. hirsutum* variety evolved through selection at JNKVV, Khandwa centre of Madhya Pradesh. The potential yield of this variety is 15–20 q/ha having moderate resistant to sucking pests, very good plant type, large boll (3.5 to 3.8 gm/boll). This variety is having shining almound brown lint colour (Figs. 1 & 2). The 2.5% span length is 23.0–25.0 mm with good strength (44.6 g/tex).

Performance of cotton culture JCC-1 for fibre, yarn and fabric properties

Coloured cotton culture developed by Khandwa Centre was also tested for fibre, yarn and fabric properties at CIRCOT, Bombay, during the year 95–96 (Figs. 3, 4, 5). This culture possessed following fibre properties: span length 24.5 mm, uniformity ratio 48 per cent, fineness 4.6 (mv), percent mature fibre 77 and tenacity at "0" gauge 36.4 g/tex. The yarn quality was found to be satisfactory at 30s count, with corrected CSP of 1929, thin places (per 100 m)–43, thick places (per 100 m)–146, Neps (per 100 m)–122, breaking strength 207 lg. Tenacity 10.6 g/tex, elongation 5.90% and ASTM Appearance–D. Six metres of fabric was also made possessing following fabric *viz.*, weave-plain, ward-30s carded, weft-30s carded ends/inch-66 and picks/inch-46. Two metre cloth has been used for stitching a shirt at CIRCOT, Bombay.

Fig. 1: JCC-1

Fig. 2: Production of JCC 1 in farmer's field

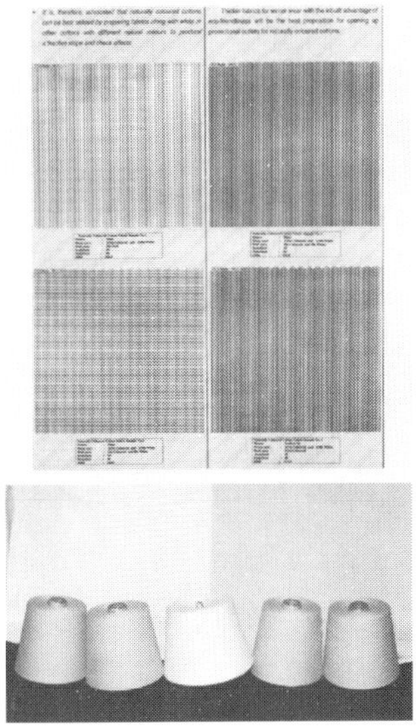

Fig. 3: Quality yarn of JCC 1 (CIRCOT)

Fabrics from JCC 1(CIRCOT)

Commercial Uses

Diploid naturally cotton has been in use for production of hand woven fabrics and this practice is being continued in some parts of coastal Andhra Pradesh. Formerly, some quantities of Cocanadas cotton was being exported. It was also used for manufacture of furnishings like curtains, mates, carpets, bed spreads and special fabrics for religious and social ceremonies.

Fig. 4: T-Shirt (Knit wears)

With better quality of coloured cotton (both diploid and tetraploid) available now, and with improvements in yarn and fabric production technology, these cottons can be used for production of a wide variety of fabrics like shirting (esp tetraploid coloured cotton), single jersey knitted fabric, double Jersey knitted fabric, denims, etc. Fabrics from naturally coloured cotton can find a niche market in view of their eco-friendly nature provided sufficient quantities can be produced at reasonable cost.

Fig. 5: Shirt

Problems and Precautions

Before their widespread adoption, a notice of caution is necessary. Growing of coloured cotton in the vicinity of white cotton is likely to contaminate the latter upsetting the seed production and cotton production process. Contamination may result from out crossing in the field and after harvesting during transportation, ginning and processing etc. Isolation at all the stages of cultivation including harvesting and post harvest processing is considered to be most crucial to prevent contamination between the coloured and white linted cottons. Growing of Desi cotton (white) around 50 metres of coloured cotton (American spp.) may offer buffer isolation. This underlines the need for suitable policy and legislative meansures for commercial cultivation of naturally coloured cotton (Shastry and Koutu, 2004). Central/State Governments have to bring forth legislation to restrict cloured cotton cultivation in identified areas only. Seasonal isolation *i.e.* growing coloured cotton in October or in non-traditional areas can help in eliminating the chances of contamination. Naturally coloured cotton production plots and seed production plots will be subject to certification for which standards have to layed down separately for natural coloured cotton grown organically and conventionally. The legislative meansures to be taken for the commercial cultivation of coloured cotton are enumerated below:

Legislative Measures (Proposed by Shastry and Koutu, 2002)

1. Under the cotton control act legislative measures for maintenance of varietal purity are extremely essential.
2. White cotton has to be protected against any type of contamination/admixture at any stage by precautions of isolation of 50 metres in fields and proper care in procurement, storage, transportation, ginning and processing etc.
3. Certain isolated cotton growing niches such as non-traditional areas of seasonal isolation may be considered. Chhindwara district of Madhya Pradesh may be a good proposition for cultivation of naturally coloured cotton.
4. The cultivation of coloured cotton has to be organized strictly on demand oriented basis in compact limited area under registration system with contract for ensured precurement and marketing.
5. Disposal of seed after ginning for oil extraction or crushing has to be ensured very strictly by the buyer under department monitoring. No seed should be allowed to flow out for sale in open market for any purpose.
6. Farmers have to be educated and guided for technical precautions, field cleaning and crop rotation to eliminate self sown seedlings.

These are very important aspects which need serious attention in our country to eliminate any chance of physical, mechanical or biological contamination at any level. It is needless to mention that care and precautions are prerequisites for utilization of such specialized technology.

1. The seed production programme for various colours and their shades under specific categories of fibre properties as per demand of textile industries should be chalked out very carefully.

2. Similarly, the production of commercial coloured cotton should be controlled under rules and regulation legislatively to ensure the quality at one hand and to prevent any misuse of the seed at the other. The seed obtained after ginning should be crushed for oil purpose and should not recycled as seed again.
3. Special arrangement for ginning, processing, procurement and marketing (export of bales/yarn/fabric/garment) should be created through a Government agency like Cotton Corporation of India or through 100% export oriented textile industries according to the world trade competition and demand.
4. The commercial production should be restricted or limited to a small number of farmers registered through an appropriate agency (like CCI, export oriented agencies etc.) ensuring the guaranted purchased of seed cotton from the farmers at remunerative prices and subsequent destruction of seed through oil extraction. The sale/purchase of coloured cotton (seed/ling/seed cotton) in the local market/mandi other than the authorized agency should be a punishable offence.
5. Cultivation of coloured cotton should be treated as a specialized field of Agricultural Technology. The ultimate success of its industrial utilization in international textile market largely depends upon the careful planning and systematic execution of the programme. Moreover, the problem and danger of environmental and water pollution through industry is also becoming a serious threat to our country. Hence in that context, the naturally coloured might be an eco-friendly gift for tomorrow.

Domestic and Export Trade Prospects

The prospects of trade of naturally coloured cotton have not been fully explored. However, looking into the concern for environment and health hazards, demand for naturally coloured cotton is bound to increase in the year to come. Concerted efforts are required on the part of marketing agencies including Cotton Corporation of India to identify buyers of naturally coloured cotton both in the country and abroad. Several mills with in the country are reportedly using organic and coloured cotton for production of yarns and fabrics. One mill in Uttar Pradesh has offered to purchase 1000 tonnes of naturally coloured cotton during 1196–97 at a premium price (Joshi, 1996).

There may be many more domestic buyers who will have to be tapped.

For export, it will be necessary to contact leading international merchants to explore the possibility of marketing this cotton. The leading buyers in Japan, South Korea and Thailand, apart from buyers in Europe, can be tapped.

It will be necessary to identify suitable area to produce naturally coloured cotton on one village one variety concept with the aim to avoid contamination with white cotton.

Marketing intelligence will have to play a crucial role in ascertaining the demand of coloured cotton at home and abroad on a continual basis so that area can be restricted accordingly; otherwise, with overproduction the grower may not get the premium on price they may be expecting for their produce.

Impact of coloured cotton cultivation on white cotton

There is already a great demand for seeds of colourd cottons from various quarters. Many of them may now possess the seeds for cultivation. It is feared that cultivation of naturally coloured cotton indiscriminately and in the vicinity of white cotton will contaminate the latter causing problems for the trade and end users of cotton. Contamination of white cottons may result from out crossing in the field and during harvest and post harvest processing. In cotton, normally 5–20 per cent natural cross pollination can be assumed though it may vary with location and extent of plant protection chemicals used (Narayanan and Sundaram, 1996). When white linted desi cotton and coloured linted uplands are grown nearby, it does not lead to out-crossing problem, but mechanical mixing of lint may occur during processing and handling unless separate processing machine are used. Transportation, marketing, ginning, pressing and spinning are main areas during processing, where physical contamination is likely to occur. Central/State Government have to bring forth legislation to restrict coloured cotton cultivation in identified areas only. There also, naturally coloured cotton production plots and seed production plots will be subject to certification for which standards have to be laid down for natural coloured cotton. When the country is likely to have commercial naturally coloured cotton cultivation immediate attention is required to be given to prevent contamination of white cotton by coloured cotton during cultivation, harvesting and post harvest processing stages. Some of the measures including policy and legislative, to be taken for cultivation, handling and processing of coloured cotton.

Quality requirement of coloured cotton

The coloured cottons have not been extensively commercialized in the past as they were too weak to be machine spun. Availability of cheap easy dyes, legislation against growing them based on the apprehension that these cottons may contaminate the white linted ones, as well as the very poor yield led to a gradual phasing out of the area occupied by these cottons. However, the recent realization of the damaging effects caused by synthetic chemical dyes on the environment, health of human beings and integration of fashion with the well-being of the ecosystem has lead to resurgence of the interest in the cultivation of the naturally coloured linted cotton all over the world (Iyer and Iyer, 1996). These cottons are considered as medical remedy for over fifty different somatic and psychosomatic disorders of man (Veerland, 1993). The quality of the fibres of the coloured linted cotton available in India is much inferior when compared to white lints. Hence, efforts are needed to improve the quality of the fibres of these cottons through various breeding methods.

References

Basu, A.K. (1996). Need for suitable policy and legislative measures for commercial cultivation of naturally coloured cotton. J. ISCI, Sept. pp. 202–206.

Christidis, B.G. and Harrison, G.J. (1955). Cotton growing problems. Pulished by McGraw Hill Book Co. Inc. N.Y., pp. 633.

Endrizzi, J.E., Turcotte, E.L. and Kohel, J. (1985). Genetics, cytology and evolution of Gossypium. *Advances in Genetics*. 23: 271–372.

Fletcher, F. (1907). Mendelian heredity in cotton. J. Agric., (2)281-282.

Fox, Sally (1987). Naturally coloured cotton- Spin-off. 46: 29–31: In search of coloured cotton- *Spin-off*. 46: 48–50.

Hutchinson, J.B. (1934). The genetics of cotton. Part X. The inheritance of leaf shape in Asiatic Gossypium. *J. Genet.*, 28: 437–513.

Hutchinson, J.B. (1935). The genetics of cotton. Part XV. The inheritance of fuzz and lintlessness and associated characters in Asiatic cottons. *J. Genet.*, 31: 4451–4470.

Iyer, K.R.K., Iyer, P.B., Sundaramurthy, V.T., Mandloi, K.C. and Meshram, L.D. (1996). Characteristics of coloured cottons and aspects requiring improvement. *J. Indian Soc. Cotton Improv.*, 21: 207–213.

Joshi, M.N. (1196). Domestic and export trade prospects of organic cotton. *ISCI J.*, (special issue on Eco-Friendly Cotton). 21(2): 234–237.

Mandloi, K.C., Khadi, B.M. and Kulkarni, V.N. (1999). Naturally colour cotton. Chapter pub. In Handbook of cotton in India. Page

Mandloi, K.C., Iyer, K.R.K., Koutu, G.K. and Julka, R. (1996). Development of coloured American cottons. *J. ISCI*. 21(2): 180–188.

Narayanaan, S.S. and Sundaram, V. (1996). Basic requirements and breeding procedures for developing naturally coloured cotton in India. *ISCI J.* (special issue on Eco-Friendly Cotton). 21(2): 159–169

Ramanathan, V. and Balasubrahmanyan, R. (1933). Lint colour in Asiatic cottons-current Science. pp. 2–128.

Shastry, P.P. and Koutu, G.K. (2002). Problems and prospects of coloured and organic cotton cultivation. *J Cotton Research and Dev.*, 2(12): 55–58.

Silow, R.A. (1944). The inheritance of lint colour in Asiatic cottons. *Journal of Genetics,* 43: 249–284.

Silow, R.A. (1944). The genetics of species development in the old world cottons. *J. Genet.*, 46: 78–115.

Sikka, and Joshi, A.B. (1960). Genetics in Cotton in India. A Monograph, Vol. 1, ICCC, Bombay, pp. 403–466.

Singh, V.V., Narayanan, S.S., Kulmethe, V. N., Anap, G.R. and Nagwekar, S.N. (1993). A brief review on cloured linted cottons ducumenting selected genotypes of *G. hirsutum* L. *ISCI J.,* 18(2): 91–94.

Singh, V.V., Punit Mohan, Kothandaraman, R. and Kulmethe, V. N. (1996). Naturally coloured upland cotton genetic resources in national gene bank. Their characteristics and utilization. *ISCI J.* (special issue on Eco-Friendly Cotton). 21(2): 151–158

Veerland, J.M. (1993). Proc. Beltwide Cotton Prod. Res. Conf. NCCA, New Orleans, USA. pp. 1535.

11

Fibre Quality

The races and varieties of the four cultivated species of *Gossypium* Linn. namely *G. barbadense*, *G. hirsutum*, *G. arboreum* and *G. herbaceum*, have characteristics fibre properties of their own. The main quality traits of cotton fibre such as ginning outturn, fibre length, fibre fineness, fibre maturity, uniformity, strength and colour have definite genetic mode of inheritance and heritability and, like other attributes, their phenotypic expression is influenced by nature of gene action, linkage, various correlation and genetic x environmental (G × E) interactions. Research investigations on genetics of these quality characters and various other biochemical studies concertinaing the degree of correlation, nature of gene action, linkage, (G × E) interaction, stability parameters and combining abilities using different biometrical models have been conducted quite intensively and extensively in various parts of the world, which have generated most valuable information for achieving desired improvement in cotton. The previous studies on the nature of gene action and association analysis between quality characters and fibre quality characteristics of popular cotton varieties and hybrids are reviewed in Tables 1–3.

Fibre Quality Evaluation

Mainly there are four fibre properties evaluated for cotton and they are useful for classification of cotton as well as ranking them for commercial use. These properties are length, fineness, maturity and strength. Conventional instruments are used for measuring cotton fiber properties, The conventional methods are very tedious and time consuming but give accurate results and hence used for research work.

In order to measure various fiber properties simultaneously and quickly, High Volume Instrument (HVI) is developed. This instrument is mainly used for testing large number of samples in short duration and comparing the results. Another instrument called Advanced Fiber Information System (AFIS) is also available for testing fiber properties particularly for small size sample. All the three type of instruments have its own merit and use in the field of cotton.

Table 1: Gene action for different fibre characters

Fibre characters	***Gene action***	***Species***	***Reference(s)***
Ginning outturn	Additive	Hirsutum	Sing *et al.*,1974; Ahmed *et al.*, 1983;
	Additive	Arboreum	Singh and Singh, 1981; Singh and Chahal, 1974; Chahal and Singh, 1975; Sandhu *et al.*, 1989; Mandloi *et al.*, 2002;
	Additive	Barbadense	Chandramati and Menon, 1973;
	Additive &	Hirsutum	Vadane, 1978;
	Dominance	Arboreum	Sandhu *et al.*, 1992;
	Additive &	Barbadense	Henry, 1974;
	Non-additive	Hirsutum	Singh and Singh, 1981; Deshpande *et al.*, 1984
Lint Index	Additive	Hirsutum	Singh and Singh, 1981;
	Dominance & Overdominance	Hirsutum	Vadane, 1978;
	Additive &	Hirsutum	Percival, 1982;
	Dominance	Arboreum	Mittal *et al.*, 1981; Sandhu *et al.*, 1992.
		Barbadense	Chandramati and Menon, 1973;
	Dominance	Hirsutum	Ramey, 1963; Ahmed *et al.*, 1983; Henry, 1974.
		Barbadense	Sandhu and Singh, 1989
	Epistasis	Arboreum	Singh and Chahal, 1974
	Nonadditive & epistasis	Arboreum	Sandhu *et al.*, 1989
	Additive & nonadditive	Arboreum	
Halo length	Additive	Hirsutm	Miller and Rawling,1967; Singh and Singh, 1981; Mandloi *et al.*, 2002
		Arboreum	Joshi *et al.*, 1969; Singh and Chahal 1974; Sandhu and Singh, 1989

Table 1: (*Contd...*)

Table 1: *(Contd...)*

Fibre characters	***Gene action***	***Species***	***Reference(s)***
		Barbadense	Henry, 1974
	Additive & Dominance	Hirsutum	Vadane, 1978; Percival, 1982
		Arboreum	Joshi *et al.*, 1969; Singh and Chahal, 1974; Singh, 1989
		Barbadense	Henry, 1974
	Dominance	Hirsutum	Ahmed *et al.*, 1983
	Partial Dominance	Hirsutum	Al-Rawi and Kohel, 1969
	Additive & Nonadditive	Hirsutum	Vadane, 1978;
		Arboreum	Tomar and Singh, 1992
Fineness	Additive	Hirsutum	Subrahmanyam *et al.*, 1991
	Additive & Dominance	Hirsutum	Nadarajan and Sree Rangasamy, 1992
	Dominance	Hirsutum	Ma *et al.*, 1983
	Additive & Nonadditive	Hirsutum	Dhanda *et al.*, 1987
		Arboreum	Chahal, 1971
Uniformity Ratio	Additive	Hirsutum	Lee *et al.*, 1967; Nadarajan and Sree Rangasamy, 1990
	Additive & Dominance	Hirsutum	Nadarajan and Sree Rangasamy, 1992
Maturity Coefficient	Additive	Hirsutum	Tabrah, 1970; Meredith *et al.*, 1970
	Additive & Dominance	Hirsutum	Nadarajan and Sree Rangasamy, 1992
	Additive & Non-additive	Hirsutum	Dhanda *et al.*, 1987; Nadarajan and Sree Rangasamy, 1990

1. *Length measurement*: Length of the cotton fibres is generally expressed as 2.5% span length or mean length or staple length. The following instruments are used for measuring length parameter.
   a. *Ball sledge sorter*: This method is based on principle of mechanically sorting fibers from a representative sliver in order of length. The fibres in each length group are collected, weighed and then with frequency distribution of fibres the mean length is determined.
   b. *Comb sorter*: In this method all the fibres of a given sample tuft are laying out side by side in order of their length beginning with longest and finishing with the shortest. The diagram so obtained on a dark surface represent distribution of fibres in

order of length. This fibre array diagram is traced on a graph paper and geometrical construction is made from which 'Effective Length' is determined. This diagram also gives the 'Short Fibre Percentage'. This method gives very reliable values and is being used for research purpose.

c. *Digital fibrograph*: The fibrograph is an optical instrument. The test sample is prepared in two combs by hand or fibro-sampler. These beards of cotton fibres are scan through light. The distance traveled by combs and light absorbed by fibres is recoded as 'length' and 'amount' on the counter. The curve obtained by plotting length vs amount is called fibrogram. 2.5% span length is the length spanned 2.5% of the total fibres caught by the comb. The uniformity ratio (UR%) is the ratio of 50% span length to 2.5% span length expressed as percentage.

2. *Fibre fineness measurement*: Fiber fineness is more commonly expressed as linear density or weight per unit length and popularly known as Micronaire. However, sometimes swollen hiar diameter or ribbon width is used to specify the fineness of cotton fibres. Direct methods and indirect methods are used to determine the fineness.

Table 2: Correlation of yield and fibre characters

Pair of characters	***Correlation***	***Reference(s)***
a. Yield of seed cotton with		
1. Ginning outturn	Significant	Arora and Sandhu, 1984; Duhoon, 1989
2. Lint yield	Significant	Waldia *et al.*, 1979; Duhoon, 1989
3. Fibre length	Negative	Sandhu *et al.*, 1986, 1987;
b. Ginning outturn with Fibre length	Non significant to negative	Duhoon, 1989
c. Lint index with		
1. Fibre length	Non significant	Stroman, 1949
2. Fibre strength	Positive	
d. Fibre length with		
1. Fibre fineness	Positive	Christidis and Harrison, 1955
2. Fibre strength	Positive	Stroman, 1949
e. Fibre fineness with strength	Negative	Singh *et al.*, 1990
f. Fibre strength with		
1. Fibre length	Positive	Singh *et al.*, 1992
2. Ginning outturn	Negative	Singh *et al.*, 1990
3. Fibre fineness	Negative	Singh *et al.*, 1990
4. MC Significant	Negative	Singh *et al.*, 1990

Table 3: Fibre quality characteristics of popular cotton varieties and hybrids

Variety	*Species*	*Year of release*	*GP*	*MFL (mm)*	*Spinning potential (counts)*
Variety					
LRA 5166	*G. hirsutum*	1982	35.0	24.5	34s
Khandwa-2	*G. hirsutum*	1971	36.5	24.0	36s
BN	*G. hirsutum*	1976	33.5	22.0	30s
F414	*G. hirsutum*	1976	33.0	22.5	30s
MCU 5 VT	*G. hirsutum*	1982	34.0	29.0	60s
G. Cot.10	*G. hirsutum*	1975	36.5	23.5	40s
DHy286	*G. hirsutum*	1975	36.0	28.0	50s
AKH 4	*G. arboreum*	1975	39.0	24.0	20s
G. Cot 11	*G. herbaceum*	1979	39.0	23.5	24s
G. Cot 12	*G. hirsutum*	1981	36.0	24.0	24s
G. Cot 13	*G. herbaceum*	1981	39.0	23.0	30s
Vikram	*G. hirsutum*	1981	36.0	22.5	40s
Khandwa–3	*G. hirsutum*	1986	35.0	23.0	30s
AKH 081	*G. hirsutum*	1987	36.0	24.5	34s
MCU5	*G. hirsutum*	1968	34.0	29.5	60s
Suvin	*G. barbadense*	1968	34.0	29.5	60s
MCU 7	*G. hirsutum*	1972	33.0	23.5	34s
J Tapti	*G. arboreum*	1996	35.0	24.5	30s
JK–4	*G. hirsutum*	1998	34.0	25.8	40s
Hybrids					
H 4	H × H	1970	34	28	50s
JKHY 1	H × H	1976	34	26	50s
H 6	H × H	1980	34	27	60s
H 8	H × H		34	26	40s
NHH 44	H × H	1983	35	24	40s
Savita	H × H	1987	34	30	60
Varalaxmi	H × B	1972	35	31	80s
JKHY 11	H × B	1981	31	30	60s
DCH 32	H × B	1981	36	33	80s
G. Cot. DH 7	h × a	1985	37	22	20s
G. Cot. DH 7	h × a	1988	34	28	40s

Source: CICR Coimbatore publication

***Indirect methods*:** The direct methods are generally tedious and time consuming. Often indirect methods are used for measurement of fibre fineness. In this method the resistance to airflow through fibre plugs is measured.

***Sheffield Micronaire*:** In this instrument, a fixed weight (3.24 g) cotton fibre plug is kept in the cylindrical chamber of the instrument and then air at constant pressure (4.7 psi) is passed the sample. The resistance to air flow measured with a float in terms of microgram per inch called Micronaire Value. For this, the scale has been calibrated on the basis of tests with us upland cottons.

3. *Fibre maturity*: Fibre maturity is an index of the extent of development of fibres or the quantity of cellulose deposition in the fibre. A cotton fiber consists of the lumen or central canal and it is surrounded by secondary wall where cellulose gets deposited. A fiber full of cellulose with negligible lumen is termed as mature fibre, where as, a fiber with negligible amount of cellulose and full of lumen is called immature fibre.

Caustic Soda Swelling Method: In this method the fibres after being swollen with 18% caustic soda are examined under the microscope with suitable magnification. Depending upon the relative dimensions of cell wall and lumen, the fibres are classified in three maturity group *viz.* mature, half-mature and immature. Around 100 to 120 fibres are examined and the percentage of each group is calculated. Using the formula

$$Mc = \frac{M + 0.6H + 0.41}{100}$$

In recent years the fibres are classified only in two groups *viz.* mature and immature and on the basis of percentage of mature fibres the cottons are classified.

For measurement of maturity, various other methods/ techniques are used such as Fibrograph, Micronaire, Polarised light method, Infra–red Spectroscopy, Dyeing method etc. However, caustic soda method is most commonly used and it is accurate and convenient.

4. *Fibre strength*: Fiber strength is a very important parameter and is generally expressed as tenacity. Breaking strength denotes the maximum tension the fiber is able to sustain before break. Dividing the breaking load by the fibre weight per unit length is termed as tenacity. Fibre strength is determined for single fibre or bundle of fibres. Single fibre strength is generally measured for research purpose, whereas, for all practical purposes fibre bundle strength is used.

For bundle strength determination there are two instruments *viz.* (1) Pressley Strength Tester and (2) Stelometer are generally used. Special uniform clamps are designed for holding the bundle of fibres for both these instruments

Other Fiber Characteristics

The length, fineness, maturity and strength are the four major physical properties of cotton fibers. However, cotton fibers do associate with other characteristics like Neps, luster, colour and trash which are very important in deciding the quality of cotton. Neps is the fiber entanglement and form a knot like structure in fibres. Trash or contamination is the forging materials present in the lint. Both these characters *viz.* Neps and Trash disturb spinning performance and effect yarn quality. Lustre and colour of cotton are also important particularly for price fixing in the market.

Ginning Outturn

Ginning is the process by which raw cotton, also called kapas, is separated into lint and seed. Present day definition for ginning machinery includes some pre ginning machines like pre cleaners, dryers, cotton feeders, lint cleaners and even the bale press, in addition to the main process of ginning itself. The ginning machinery should be set to perform efficiently and at the same time to cause minimal effect on the quality of both the seed as well as the lint.

Arboreum varieties generally have small seeds and several of them yield short stapled cottons with 2.5% span length ranging from 17.5 to 25 mm. The cottons, after ginning yields fuzzy seeds *i.e.* the seeds possess short fibres. Most of the arboreum varieties are hard to gin. They give high ginning percentage ranging from 35 to 45% *i.e.* amount of lint (expressed as a percentage) per unit weight of kapas is very high. The cottons are, generally, coarse.

Herbaceum cottons also give small seeds and ginned seeds appear naked. But ginning is much easier. The seeds are slightly grayish. The original cotton plants are perennial and this feature distinguished them from other varieties. Most of these cottons are finer as compared to Arboreums. Very little variation is found in their properties. The important variety of this group is Digvijay.

Table 4: Classification of cotton on the basis of fibre properties

Length	
Range of 2.5% span length	***Category***
20 mm and below	Short
20.5 mm–24.5 mm	Medium
25.0 mm–27.0 mm	Medium long
27.5 mm–32.0 mm	Long
32.5 mm and above	Extra long
Ratio Uniformity	
Range of UR%	***Category***
Below 42	Poor
42–43	Fair
44–45	Average
46–47	Good
Above 47	Excellent
Micronaire	
Range of Micronaire	***Category***
Below 3.0	Very fine
3.0–3.9	Fine
4.0–4.9	Average
5.0–5.9	Coarse
6.0 and above	Very coarse
Strength	
Range of fibre bundle strength (g/t) at 3.2 mm gauge	***Category***
16 and below	Very low
17–20	Low
21–24	Average
25–28	Good
29 and above	Very good

Source: CIRCOT

In hirsutum cottons, the seeds are fuzzy and as these fuzz fibres are small in length they are left on the seed surface after ginning. The seeds are generally large and irregularly shaped. While some of them are easy to gin some others are quite hard to gin. The staple length also varies between varieties from 21.5 to 29.5 mm. This group occupies the largest acreage under cotton cultivation in India. The seeds can be classified on the basis of their size and also on the basis of the amount of fuzz left on them after ginning. The ginning percentage shows a very high variation and there are some commercial varieties showing 42% ginning percentage.

The *G. barbadense* cotton species also known as Egyption cotton and are the finest cottons among all the cultivated species. The seed obtained after ginning is naked *i.e.* without fuzz and are greenish black in colour. The ginning percentage is less then 30% in general.

The cotton trading is based on the grade and the ginning percentage of the cotton. The grade of the cotton is known by its staple length, whiteness (colour) and the ginning percentage, while the ginning percentage is judged by the amount of lint obtained from the given quantity of kapas.

Ginning percentage is nothing but the amount of lint yield from the given quantity of kapas. If the lint yielded is higher, the G.P. will be definitely higher. G.P. depends upon the genotype of cotton, its seed size and weight, micronaire and the lint density,

$$\text{G.P.} = \frac{\text{Weight of lint}}{\text{Weight of seed cotton} \times 100}$$

Table 5: Characteristics affecting the G.P.

Characteristic/ Genotype	***G. arboreum***	***G. herbaceum***	***G. barbadense***	***G. hirsutum***
Size of the seed	Small to very small	Medium to small	Large to very large	Large to very large
Staple length	Small to medium	Small to long	Long to very long	Medium to long
Micronaire (Fineness)	High, coarser	Coarse to fine	Low, fine	Coarse to fine
Fuzziness of seed	None	Less	None	Moderate to high
G.P.	High	Moderate	Low	Moderate to high
%Short fibre contents	4.0	2,7	1.5	1.7

Ginning machine

Fibres are separated from the seed coat during the process of ginning. Whether it is roller ginning or saw ginning, fibres do undergo physical stress that causes the breakage, this is the reason for the reduction in the fibre.

Roller Gins

The roller gins are classified as SR (single roller) or DR (Double roller) depending on the number or rollers per machine. According to a recent survey of the ginning industry in India, there are 28493 SR gins and 19126 DR gins making a total of 47619 and these are spread over 3200 gin factories.

Saw Gins

The output of the saw gins depends on the number of saws, number of teeth per saw, the sharpness and angle of the teeth, speed of the rotating shaft, air blast pressure and the condition of the cotton. The lint productivity is 300 to 400kg/hr. Saw ginned lint is more regular and uniform. Saw ginned lint is more regular and uniform and blow room losses are comparatively less.

Double Roller Gin

Double roller gins are being extensively used in India to gin cotton. The amount of lint produced per hour per gin is called Ginning Out Turn (GOT) and at present varies between 35 to 45 kg. The two most important machine parameters that influence the productivity of the lint in DR gins are the speed of the roller and oscillation frequency of the beater. Almost all the commercial double roller gins in ginning industry employ a fixes roller speed of 90–100 rpm, while the beater itself oscillates with a frequency of 900–1000 rpm, the ratio of these speeds being 1:10 and that was suitable for the old world cottons having staple length about 22 to 24 mm.

Table 6: Difference between Roller gin and Saw gin

Double roller gin	***Saw gin***
No in built pre-cleaners are with gin stand. The mobile cleaners keep out side near the heap. The lint after ginning is trashy	Pre–cleaning units such as inclined per-cleaner. Extractors and super lint cleaners are in built in the saw gin stand so lint is clean.
The DR gin has two rollers of 40 inch length and runs with 100 rpm speed. The fibres are gripped between roller and fixed knife and pulled. It is a gentle ginning. Easy to operate and less maintenance cost. Power consumption is also less.	It contains 75 to 120 circular saws of 9 inch de. Equally placed along the length of the shaft, runs with the speed of 700 rpm. Fibres are caught in the saw-teeth and pulled with great force. It is a harsh ginning. Difficult to handle and high maintenance cost
The lint productivity of DR gin is 40–50 kg/hr	The lint productivity is 300 to 400 kg/hr
It gins almost all type of fibres giving high G.P.	It is a selective ginning. Gins longer length of fibres hence G.P is less
Less fuzz remains of the seed.	Fuzz proportion is more on seed
Roller ginned lint is less neppy	Saw ginned lint is more neppy
The lint is irregular and so blow room losses are more	Saw ginned lint is more regular and uniform, blow room losses are less
Suitable for all types of cottons	Not suitable for ginning extra long staple cottons.

Trash in Cottons

Cotton, in general, contains heavy and fine trash, any material other then cotton itself is termed as 'Trash'. Trashy lint affects the quality of the yarn and the fabrics made from it. Trash also affects the performance of the gin machinery and is the cause for the wear and tear of the same. Thus the study of trash assumes importance. The trash content in cotton is one major source of worry for the spinning industry. Nowadays, the trash components are referred to as contaminants.

Pre-cleaning machines

Cotton has to be free from all sorts of trash before it is ginned. The machines that perform this vital function are generally named as cleaning machines. Trash is classified as heavy or light trash as explained earlier. The lighter trash is removed with the aid of 'Cylinder Cleaners' whereas heavy trash is tackled by 'Extractor

Cleaners'. Nowadays the ward pre-cleaner is associated with the former class of cleaner *viz.* Cylinder Cleaners alone.

Some important points

1. Short staple cottons spun to lower count while long staple cotton spun to higher count.
2. Yarn spun form fine fibres will be more even have large number of fibres in cross section and strong compared to yarn spun from coarse fibres (same count).
3. Immature fibres causes lapping of fibres on rollers, lowering yarn appearance grade due to neps formation, uneven dying.
4. Immature fibres show lower weight per unit length than the mature fibres in the same cotton hence affect the fineness.
5. Fibre strength is the major factor influences the strength of yarn.
6. Neps and trash contents in cotton play crucial role in deciding spinning performance and yarn quality *viz.,* imperfections.

References

Ahmed, A.K., Al-Rawi, M. and Harban, S.R. (1983). *Iraqui J. of Agri. Sci.,* 1: 133–138.

Al-Rawi, K.M. and Kohel, R.J. (1969). *Crop Sci.,* 9: 779–83.

Arora, R.L. and Sandhu, B.S. (1984). *Cotton Dev.,* 14 (2-3): 19–21.

Chahal, G.S. (1971). M. Sc. Thesis PAU Ludhiana (unpublished).

Chahal, G.S. and Singh, T.H. (1975). *Crop Improv.,* 2: 25–35.

Chandramathi, P.S. and Menon, P.M. (1973). *Madras Agric. J.,* 66: 1950–66.

Christidis, B.G. and Harrison, G.J. "Cotton growing problems" Pub. By McGraw Hill Book Co. NY, Edn. p. 633.

Deshpande, L.A., Vyahalkar, G.R., Ansingkar, A.S. and Mane, S.S. (1984). *Indian J. Agric. Sci.,* 54: 97–99.

Dhanda, S.S., Tyagi, A.P., Singh, D.P. and Rana, O.P.S. (1987). *J. Indian Soc. Cotton Improv.,* 12(2): 84–87.

Duhoon, S.S. (1989). *J. Indian Soc. Cotton Improv.,* 14(1): 39–44.

Henry, S. (1974). M. Sc. Thesis, IARI, New Delhi.

Joshi, A.B., Singh, Munshi and Kadapa, S.N. (1969). *Indian Soc. Agric. Stat.,* pp. 269–278.

Lee, J.A., Miler, P.A. and Rawling, J.O. (1967). *Crop Sci.,* 7: 477–481.

Ma, F.C., Zhou, Y.Y., Wang, R.T. and Lin, C.G. Acta Agriculture University.

Mandloi, K.C., Koutu, G.K. and Mishra, U.S. (1996). *J. Indian Soc. Cotton Improv.,* (1): 5–14.

Mandloi, K.C., Koutu, G.K. and Mishra, U.S. (1998). *J. Indian Soc. Cotton Improv.,* (2): 147–152.

Miller, P.A. and Rawling, J.G. (1971). *Crop Sci.*, 11: 695–697.
Nadarajan, N. and Sree Rangasamy, S.R. (1990). *J. Indian Soc. Cotton Improv.,* 15(1): 16–19.
Nadarajan, N. and Sree Rangasamy, S.R. (1992). *Indian J. Genet.*, 52(3): 245–251.
Percival, A.E. (1982). *Diss. Abstr. Int.,* 43: 627.
Pakinensi. 9(4): 27–34, (1983). Cotton Trop. Fibres. *Crop Abstr.*, 9(5): 210.
Ramey, H.H. (1963). *Crop Sci.*, 3: 32–33.
Sandhu, B.S., Gill, M.S. and Mittal, V.P. (1989). Cotton Dev., p. 18.
Sandhu, B.S. and Singh, R.P. (1989). *Punjab Agriculture University Research J.*, 26: 10–13.
Singh, M., Singh, V.P., Lal, C.B. and Paul, K. (1990). *Indian J. Agric. Sci.*, 602: 137–8.
Singh, M., Singh, V.P. and Patil, N.B. (1992). *Indian J. Agric. Sci.*, 62(5): 325–8.
Singh, P. and Singh, H.G. (1981). *Indian J. Agric. Sci.*, 51: 209–13.
Singh, T.H. and Chahal, S. (1974). *Indian J. Genet.,* 34: 323–25.
Singh, T.H., Kandola, H.S. and Nagi, P.S. (1974). *Indian J. Agric. Sci.*, 44: 521–24.
Stroman, G.N. (1949). *J. Agric. Res.,* 78: 353–364.
Subrahmnyam, Y., Patel, J.C., Mehta, N.P., Maiurla, A.T., Bhalod, M.C. and Kanaki Iyer, K. (1991). *Indian J. Genet.*, 51(4): 424–428.
Tabrah, T.A. (1970). Ph. D. Thesis. Okalhoma State Uni. USA.
Tomar, S.K. and Singh, S.P. (1992). *Indian J. Genet.*, 52(3): 264–269.
Vadane, N.N. (1978). M.SC thesis MPAU. Rahuri. pp. 413–722.
Waldia, R.S., Jatasra, D.S. and Dahiya, B.N. (1980). *Indian J. Agric., Sci.* 50: 899–901.

12

Cotton By-Product Utilization

Utilization of Cotton Seed

Cotton seed is an important article of food in India for milch cows, buffaloes and working animals. Before commercialization of the milling of cotton seed for oil, all the cotton seed production in India was being utilized as cattle feed, after meeting the requirements for planting, for which about five percent of the seed was utilized. Presently with popularization hybrid technology with *Bt.* background, the entire cotton seed after ginning is only used for oil extraction and cake is used for animal feed. The major constituents of cotton seed are fat, protein, moisture, crude fibre and carbohydrates. The minor constituents are pigments, phosphorous compounds, anti–oxidants, mineral compounds and other substances. The composition of cotton seed varies within certain limits, with the species, variety, maturity, condition of cultivation and environment.

The dark colour of extracted cotton oil requires refining. The colour of cotton seed oil is mainly due to the presence in it of gossypol and gossypol like pigments. These are extremely unstable and sensitive to oxidation. With the modern techniques for refining of cotton seed oil for gossypol contents, the present day industries are utilizing maximum potential of oil used for human consumption as Vanaspati.

Cotton seed which forms about 2/3 of seed cotton is a valuable by-product of cotton crop containing 15 to 23 per cent edible oil. Cotton seed oil is rich in essential fatty acids which is the most important one, is present to the extent of 51%. The keeping quality of the oil is comparable with that of other important oil seed crops. The crude oil is used for soap making, lubrication and lightning in villages. With the present level of availability of cotton seeds (>30 lakh tonnes) there is a potential for production of 5.5 lakh tonnes of cotton seed oil. However only 3 lakh tonnes per annum cotton seed oil is produced. The unused cotton seed is manly used for cattle feed and partly to non–scientific process of extraction of oil. Educating the farmers to feed only cotton seed oil cake which gives the required nutrient to the cattle, instead of whole seed will divert more cotton seed for oil extraction. Hence, in year ahead, cotton seed oil will play an important role in meeting the demand of edible oil in the country.

Table 1: Composition of cultivated species of cotton

Seed component	***G. arboreum***	***G. herbaceum***	***G. hirsutum***	***G. barbadense***
Seed oil (%)	14.6–22.1	16.9–22.4	16.1–24.4	17.9–25.6
Protein (%)	28.3–38.1	30.7–37.9	30.3–42.8	32.0–41.1
Free gossypol content (%)	0.30–1.25	0.43–1.09	0.42–1.25	0.73–1.49
Linters (%)	3.9–11.5	2.6–7.8	3.1–15.5	–
Kernel (%)	42.0–57.8	48.5–57.3	46.5–60.3	59.6–63.4

Source: Compiled from cotton monograph

Gossypol

Gossypol is one of the naturally occurring substances in all species of cotton. It is a polyphenolic yellow terpenoid aldehyde pigment chemically known as 1,1',6,6'7,7'–hexahydroxy–5,5'–iisopropyl–3,3'–dimethyl (2,2'–binaphalene) 8,8'–dicarboxaldehyde, which is toxic to certain animals, particularly poultry, and swine (Harms and Holley, 1951). It is present within discrete bodies (sub–epidermal glands) which are known as gossypol glands or pigment glands, anatomically peculiar to leaves, stems, squares, boll, roots and seeds of cotton plant. Pigment gland is one of the important key characters used to define the tribe of the family Malvaceae. The tribe Gossypieae includes the genera *Gossypium* along with seven other genera (Fryxell, 1968). Thus the plant of genus *Gossypium* contains several pigments unique to the plant kingdom and gossypol is one of them. The purpose of the glands has been a matter of speculation as Cook (1906) suggested that they may be a protective mechanisms against insects. However the toxicity of gossypol has been well documented (Balley, 1948). However, the toxicity of gossypol to *Heliothis* species was demonstrated by Lukefahr and Martin (1966). Thus the probable role of gossypol in conferring bollworm resistance in cotton was established. Later on very extensive work on the role of gossypol glands in relation to insects resistance in cotton has been done throughout the world.

The gossypol glands of cotton mainly contain three compounds namely gossypol, which is a terpenoid aldehyde, terpenoid quinines and their helicides derivatives and imine *i.e.* gossyrubiline-4. Gossypol (2a) and its methyl ethers 2b, 2c are more predominant *i.e.* 90–95% in seed embryo, stem, phloem, root, petals and other parts of the plant which are achlorophyllus and shielded from sunlight. Gossypol glands in green tissues contain predominantly the terpenoid quinines and their heliocides derivatives.

There is a every scope of developing cotton varieties containing pigment glands on the plant body but not in the seed cotyledons or endosperm so that it can be utilized easily for edible oil extraction purpose, minimizing the cost of removal of gossypol in refining process. The research findings of Bell *et al.* (1977) and Lukefahr *et al.* (1977) have already established the synthesis of new recombination by interspecies breeding between *G. hirsutum* and *G. barbadense* containing different type of gossypol compounds indicating the possibility to develop cotton varieties with these chemicals for maximum inbuilt protection against the bollworm on one hand and glandless cotton seed on the other (Koutu *et al.*, 1994).

Production of Biogas and Compost from Ginnery Waste

Ginning is the first mechanical operation that the seed cotton undergoes during which the lint is separated from seed, Indian cottons fetch lesser price even in the domestic market as they contain large amounts of trash in baled lint. Foreign cottons being less trashy and contaminant free, therefore Indian mills are resorting to cotton imports. This definitely affects the price of cotton. The sincere efforts of Govt. of India to modernize Market Yards and ginning and pressing industries under Mini Mission III and IV of Technology Mission on Cotton have already started showing positive results in producing clean bales par with International Standards.

Due to the introduction of pre and post cleaners coupled with other best management practices of machine handling cotton produces different kinds of unspinnable cotton wastes and trash in modernized ginning and pressing factories. These units need not be alarmed by the increased waste generated during processing. In fact, the following are the advantages.

A. Modernised G & P factories are producing clean cotton which can definitely fetch better price.
B. The loss incurred during the process *i.e.* short fibres and trash can be put to better value addition in terms of use of this waste material for compost.

Table 2: Proximate chemical composition of Willow–dust and Cattle waste

Constituents	*Willow – dust (%)*	*Cattle waste (%)*
Ash (including sand)	10–12	15–18
Cellulose	24–28	27–29
Hemicellulose	15–18	25–26
Lignin	14–16	16–18
Nitrogen	1.0–1.5	1.0–1.2
Ether extractives	4–6	3-4
Water solubles	22–25	9–10
C/N	28–30	30–32

Mushrooms production on Cotton Plant Stalks

The edible mushrooms are delicacy in food and form one of the choicest table dishes. They are not staple food as such but can serve as an important supplement to diet. They are rich in protein and an excellent source of vitamins and minerals. Most of the mushrooms have very low starch content and can form an ideal food for diabetic patients. Though, about 2000 species have been reported to be edible, the techniques of artificial cultivation are available only for a dozen of species of which about four are being cultivated on a commercial scale.

It has been well established that oyster mushrooms can be artificial cultivated on various lignocellulosic crop residues, *viz.*, rice straw, wheat straw and cotton straw.

	Mushroom species			
Substrates	*P. sajor-caju*	*P. florida*	*P. flabellatus*	*P. ostreatus*
Wheat straw	400	410	417	440
Rice straw	310	425	307	453
Bajra straw	365	440	362	513
Cotton stalks	480	453	450	600

Mushroom Species

Production of Particle Boards and Hard Boards from Cotton Stalks

Cotton farmers can be benefited with the use of cotton stalks for preparation of particle and hard board. This alternative means for increasing the returns from cotton farming by judicious use of byproducts from cotton cultivation are to be explored which will generate additional sources of income to the farmer.

It is estimated that about 15 million tonnes of cotton plant stalk is generated in our country every year. Most of the stalk produced is treated as waste though, to some extent, it is being used as fuel by the rural masses, while at the same time bulk of the stalk is burnt off in the fields after the harvest of the cotton crop. Cotton plant stalk contains about 60% holocellulose, 27% lignin and 7% ash. In contrast to other agricultural crop residues, cotton stalk is comparable to the most common species of hardwood in respect of fibrous structure and hence it can be used for the manufacture of particle boards, preparation of pulp and paper, hard boards, corrugated boards and boxes, microcrystallie cellulose, cellulose derivatives, for growing edible mushrooms, etc.

Particle Board is panel product produced by compressing small particles of wood while simultaneously bonding them with an adhesive. The various types of particle boards differ greatly with regard to the size and geometry to the particle, the amount of resin (adhesive) used and the density to which the panel is pressed. The properties and potential uses of board differ with these variables. Particle board has emerged as a versatile substitute for wood in many applications such as paneling, partitioning, false ceiling, furniture making, etc, In building construction, particle board finds its application for decorative and acoustic purposes, if AC duct covering, door panel inserts, wall panels, floor, ceiling, tiles, etc. In transportation sector, particle board is used in the railways for sliding doors whereas it has also use in making "Printing Block".

The production of particle boards shows an average growth rate of 15.5% per annum during last decade as against wood products growth rate of 70% per annum and of plywood at 5% per annum. Considering the past growth rate, usage pattern, the actual demand for particle boards is expected to be around 95,000 tonnes, where as the supply position is not even 60% of the demand. Hence there is a good market potential for particle board in our country.

Advantages of particle board are

- It is free from natural defects of wood, like warping.
- It is easy to fix. For instance, the factory made panel doors with particle board are available in ready to fix form. Similarly, for wall paneling, false ceilings, table tops etc.
- It is cheaper than substitute materials.
- With proper protective surface coating and edge covering, particle board can be made water resistant, termite proof and fire resistant. It can take a variety of surface finishes, like laminations, veneers, paint, varnish polish, etc, Attractive wall paper can also be used as surface finish.

Cotton seed cake for livestock and poultry feed

The value of protein in the dairy feed of animals had been long recognized. In addition to supplying protein, the cotton seed cake is a source of phosphorous and fat. For dairy cows, the cotton seed cake furnished the most important and cheapest source of protein .Cotton seed cake and meal are standard protein supplements for the feeding of sheep, being extensively used maintenance of range and farm flocks. An increasing amount of cotton seed meal is being used as a source of protein in feed of growing chickens. It supplies very effectively the added protein necessary for satisfactory growth when fed in a mash containing dried butter milk.

Cotton Seed Flour

***Edible product*:** As cotton seed is a large potential source of high protein foodstuffs, for which there is somewhat of a shortage generally in the world. As it is a rich source of certain vitamins and essential amino acids in which wheat and other cereals are naturally deficient, the possibility of utilizing cotton seed on large scale as a source of food for human beings is at present very useful to nourish the poor rural people in future.

***Linters*:** Linter is next important to oil and cake among the cotton seed products. The amount of linters left on the seed as it comes from the gin varies considerably from 1 to 17.9 per cent. The variation in the range due to the variety of cotton and to environments, including weather conditions, rather than the method of ginning. The amount of linters removed from the ginned cotton seed varies according to amount of total linters remaining on the seed market value of linters and efficiency of oil mills.

Clean cotton linters are almost pure (88 to 96 per cent) cellulose, but linters, as they leave by the oil mills, contain many kinds of foreign matter in varying amounts which reduce the cellulose content. Cotton linters probably have a wider variety of end-use markets than any other cotton seed product. Small amount of first cut linters primarily grade one and two containing the long fibres, are used in the manufacture of medical supplies, such as surgical dressings, absorbent cotton and swabs (after bleaching and sterilizing the linters), twin, yarns mops and wicks. The remaining first-cut linters are used primarily in the batting, padding and felting trade such as automobile and furniture padding and bedding.

***Hulls*:** Cotton seed hulls are the least valuable of the cotton products, contributing only about three per cent. Originally, the hulls were used primarily as fuel for mills operation. The hulls are also used as a stuffing and packing material and soil conditioner. Though cotton seed hulls are being used to a limited extent as roughage, probably due to the transport cost of this bulky material, a major part of them is used at present as fuel. However their use as fuel is not economical. Therefore it is desirable to popularize the hull for use in chemical industries.

References

Bailey, A.E. (1984). "Cotton sed and cotton seed products". Interscience Publisher, Inc., N.Y. Edn. pp. 213–32.

Bell, A.A. and Stipanovic, R.D. (1977). Proc. Beltwide Cotton P*rod. Res. Conf. Jan.* 10–12, Atlanta, GA. p. 244.

Cook, O.F. (1906). U.S. Dept. *Agric. Bur., Plant Ind. Bull.*, No. 88.

Fryxell, P.A. (1968). *Bot. Gaz.* pp. 129–296.

Harms, W.S. and Holley, K.t. (1951). *Proc. Soc. Exp. Biol. Med.,* 77: 297–299.

Lukefahr, M.J. and Houghtaling, J.E. (1969). *J. Econ. Entomol*. 59: 176–179.

Lukefahr, M.J., Stipnovic, R.D., Bell, A.A. and Grey, J.R. (1977). Proc. Beltwide Cotton Prod. Res. Conf. Jan. 10–12. Atlanta, GA. 97.

Koutu, G.K., Mandloi, K.C., Mishra, U.S. and Singh, C.B. (1994). *J. Indian Soc. Cott. Improv.*, 19(2): 124–138.

13

New Wilt of Cotton

New Wilt of cotton is a challenging problem for plant scientists from all the disciplines. A peculiar wilt like disease was observed in 1978, for the first time in India simultaneously from Adilabad (A.P.) and Khandwa (M.P.) by Mandloi (1979) on the widely cultivated hybrid JKHy-1. Since then it has been reported to occur in moderate to severe form on the other varieties and hybrids in the other parts of the country also. Today, even after twenty five years of its occurrence our knowledge about the malady is fragmentary. The problem has been in the limelight again with the introduction of *Bt* cotton which seems to be comparatively more susceptible to New Wilt as compared to other varieties and hybrid. An attempt has been made to compile the available information on various aspects and presented in the form of review.

Occurrence

The disease was reported from Maharashtra in 1978 from Deccan Canal Area on Varalaxmi in the incidence of 8–10% , later assuming epiphytotic proportions (20–80%) in 1983 on DCH–32, MECH–1, H–6 and RHH–195 (Ekbote, 1984) and in 1979 from Marathwada ranging between 1–10% in 1982 and 1984 (Mali *et al.*, 1984). Bhale (1984) reported the disease from Maharashtra and Tamil Nadu in 1981 on DCH-32. It assumed serious proportions in Madhya Pradesh and Maharashtra between 1982 and 1984 on MECH–1 (Mandloi *et al.*, 1984; Raj *et al.*, 1984) and in Tamil Nadu (Srinivasan, 1984). Later reports from Andhra Pradesh between 1980 and 1984 indicated an existence of 5–50% (Reddy *et al.*, 1984).

Extensive surveys carried out in M.P. between 1984 and 1987 indicated its severe outbreaks in 1984, 1986 and 1987 on JKHy-1, H-4, H-6 and MECH-1 (Mandloi *et al.*, 1984; Anonymous, 1985a, b, 1986, 1987). Surveys in 1989 and 1990 showed an average incidence between 3.0 and 6.2% (Shastry, 1993). Its first report from Gujarat came in 1979 on DCH-32 (Mehta, 1984). Later moderate to severe incidence was observed in 1984 and sporadic occurrence in 1986 (Mehta, 1987).

The wilt was reported on Somnath in Haryana in 1986 and later on with an incidence of about 10% (Raj *et al.*, 1987). Some cases of sudden wilting were observed in 1988 in Rajasthan (Anonymous, 1989).

With the introduction of *Bt* hybrids in 2002–03, reports of large scale sudden drying have been reported from Madhya Pradesh, Maharashtra and Gujarat. In M.P. the problem was particularly severe on *Bt* hybrids in 2002–03 and 2005–06 seasons.

Loss–Seed Cotton Yield/Fibre quality

Wilt is responsible to cause losses in quantity as well as quality on cotton. Losses in yield to the tune of 37.92 and 48.9% have been reported on varieties Laxmi and Varalaxmi, respectively (Anonymous, 1985a, b). Loss in seed cotton was shown as 15.0 and 14.28% on NHH-44 and NHB-12, respectively (Anonymous, 1990). Raj *et al.* (1990 a, b) observed that a 15.64 to 67.77% reduction in number of bolls corresponded to the 44.75 to 83.54% reduction in yield on a 3–4 months old crop. Similarly, Mayee *et al.* (1991) observed 49–65% reduction in seed cotton yield due to wilt at different stages of growth. In germplasm accessions 28.9 to 59.6% and hybrids 33.6 to 60.5% reduction in seed cotton yield has been observed. The wilt affected plants show considerable reduction in seed and lint indices, boll weight and seed oil content (Narayanan *et al.*, 1989). The terminal losses as reduction in number of bolls, seed cotton yield on plant basis have been shown to be 43.96 and 44.99% in wilted and 36.58 and 34.58% in wilt recovered plants. The yield losses on field basis were reported to be 1.3–4.37% (Shastry, 1993). Wilt is also responsible for reduced seed germination (Mayee *et al.*, 1991; Shastry, 1993). Wilt is responsible for reduction in mean fibre length, fineness and maturity (Mandloi *et al.*, 1987; Deshmukh *et al.*, 1989). It was shown that the worst effect of wilt was on fibre maturity (Raj *et al.*, 1990a, b; Shastry, 1993). The lint from wilt affected plants contains high percentage of linters (Jadhav *et al.*, 1992).

Symptomatology

It has been consistently been observed that this disorder remarkably appears at the boll development stage. According to Basu (1984), the wilting is seen in vegetative, square, flowering and early boll formation stages of the crop, the first symptoms appearing when the crop is five to six weeks old. Mali *et al.* (1984) observed the symptom to appear when the plants were 55–65 days old. Ekbote (1984) and Shastry (1993) have recorded the influence of date of sowing on wilt and found that irrespective of the stage of the crop, wilting occurred in a short span of period. The crop stages, when wilting occurred, were 14–16, 12–14 and 10–12 weeks respectively. Balsubramanya *et al.* (1992) found that wilt occurred at the early boll development stage. At Adilabad and Nagpur, the disease was found to appear at flowering and boll development stages (Srinivasan, 1981, 1984; Raj *et al.*, 1984; Bhale, 1984). Jadhav *et al.* (1992) observed maximum impact of adverse climate in the form of wilt at linting stage. The higher susceptibility of *Bt* hybrid is being linked to the higher boll retention due to efficient bollworm control.

Development of the wilt was closely watched by several workers from initiation of the disease to the stages documenting the exact syndrome. The major steps have been categorized after critically discussing the available picture. Various stages of symptoms of this wilt clearly show that it is different from earlier known pathogenic wilts caused by *Fusarium* and *Verticillium* (Ebbels, 1975). Characteristic symptoms of new wilt have been described by Mandloi *et al.* (1985) as:

- Sudden drooping of leaves, giving a morbid look to the plant.
- Bronzed green lean surface, reddened petiole and broadening of leaf angle.
- No necrosis and drying of drooped leaves.
- Disorder found as random scattered plants commonly one plant shows wilt out of two at the same hills.
- The root system remains healthy without any internal discolouration.

- The wilted plants revive spontaneously after a pause of 4-6 weeks.

The wilt symptoms were categorized into two groups *i.e.* the slow wilt recognized at an early stage where the leaves become morbid but do not loose their turgidity immediately, turn yellow, reddening starts from margins extending to petiole and stem thus requiring 12–15 days to express complete symptoms; the quick wilt, on the other hand, is typified by sudden drooping of leaves at peak flowering and bolling stage, rapid drying of leaves wilts out pigmentation taking 7–10 days in wilting (Raj *et al.*, 1987, 1990 a, b, 1991). This has been named as parawilt (Mayee *et al.*, 1989) and Adilabad wilt.

These typical symptoms can be distinguished from the similar type of earlier reported wilts by comparing the critical stages described for them as shown in Table 1 (Ebbels, 1975). The disorder has been shown to occur under stress (water stress as well as water-logged) conditions (Shastry, 1993). Forced bursting of immature bolls is also observed due to wilt. The disease is prominently seen at flowering and boll development stages of the crop. Most of the workers are of the opinion that one plant out of the two at a hills gets affected in this wilt; however, in rare cases both the plants at a hills show wilt (Mandloi *et al.*, 1984; Srinivasan, 1984; Raj *et al.*, 1984, 1990a, b, 1991; Shastry, 1993). This has been explained by Shastry (1993) on the basis of probability estimates. The average wilt incidence varies around 4–6% *i.e.* any 4–6 plants at 50 dibling points will wilt with probability P=0.0768–0.1128. Further the probability of any two plants (out of these 4–6) showing wilt at the same dibbling point is still lower as P=0.0016–0.0036. This probability will naturally be higher as the incidence level increase.

Several biotic and abiotic agents have been proposed as probable cause of this malady from time to time. These are being discussed herein with possible evidences in their support.

Biotic agents

The rhizosphere of the affected plants have often been found to be associated with higher nematode populations. Laxamanan *et al.* (1989) indicated that the disorder could be due to infection by *Rotylenchulus reniformis*. Kurundkar and his associates carried out systematic studies on this aspect for three years and observed that

Table 1: A comparison of symptoms of earlier known wilts with New Wilt at different stages

Stage	***Characteristic symptoms of***			
	Fusarium	***Verticillium***	***Lubaga***	***Wilters***
Occurrence	Usually in patches, may be large, symptoms becoming less severe towards periphery, appears early, favoured by sandy soils, root knots and warm weather.	As small group plants, appears late in the season, favoured by alkaline soils	As single or small patches with more severity in the centre. and cool night.	As randomly scattered single plants.
Stem	Vascular tissue stained brown, especially immediately beneath the bark.	Vascular tissue stained brown usually in scattered bundles through the wood	Vascular tissue stained brown as in Fusarium.	Vascular tissue unstained, stem becomes reddish externally.
Leaves	Yellow, often parts of the leaves wilt, shrivel, become brown and fall down.	Yellow at edges but do not fall.	Yellowing without interveinal yellowing and necrosis	Become bronzed green, petiole reddish and widen the angle with the stem.
Growth	Ceases, infection in young plants fatal, new growth from lower nodes.	Retarded.	Ceases.	Ceases or much retarded for 2-6 weeks, no new growth from lower nodes.
Cause	Pathogenic.	Pathogenic.	Non-pathogenic, but sometimes yield *Fusarium equisetii.*	Non–pathogenic, partly physiological, partly genetic.

Shastry (1993)

the nematodes reduced the growth of the plants but did not produce the symptoms of New Wilt (Mayee, 2001).

Mali *et al.* (1984) suggested the need to study the involvement of flagellates in this disorder. Mayee (1993) confirmed the association of flagellates like organism with the phloem tissues of wilts affected plants. However, it has not been possible to typify the nucleus, kinetoplast and flagellum for clearly establishing the presence of protozoal structures. Indirect evidence of protozoal etiology has also been obtained when temporary remission of wilt symptoms has been possible by spraying metonidazole coupled with carbendazm which are known antiprotozoal drugs.

Repeated attempts to isolate a pathogen from the roots of wilted plants of various arboreum and hirsutum varieties and hybrids were not successful to establish fungi, bacteria and actinomycetes as a pathogen associated with the wilt (Mandloi *et al.*, 1984, 1985; Shastry, 1993). Although 10% of the samples yielded Fusarium, Mali *et al.* (1984) failed to induce typical symptoms. They have isolated a yellow bacterium on nutrient agar from wilting plants; however, its nature as an etiological agent could not be established.

Involvement of viruses has been ruled out by many investigators since the disease could not be transmitted by grafting *in situ* (Anonymous, 1987a; Krishna *et al.*, 1988; Mali *et al.*, 1984; Mayee *et al.*, 1991; Raj *et al.*, 1991; Shastry, 1993) and through insects (Jayaswal, 1984; Mayee *et al.*, 1991). The seeds used from previously wilted plants showed poor and inconsistent recurrence of wilt (Anonymous, 1986). Similarly, the recurrence of wilt in the progeny grown from seeds of wilt affected plants was found erratic (Mayee *et al.*, 1991; Shastry, 1993).

Scanning Electron Microscopy

Vascular tissues of wilting plants remain healthy and unstained (Mandloi *et al.*, 1985). Roots from affected plants do not show any discolouration or clogging (Balasubramanaya *et al.*, 1992). (Mayee *et al.*, 1991) through SEM studies have revealed the presence of oval and irregular bodies in the phloem cells of root sections from wilted plants. They considered them to be accumulation of nutrients under moisture stress. Studies of microtome sections of roots, stem and petioles under light microscope (400 X) reveal that there is no apparent blockage of xylem vessels as is known to occur in other wilts. There is no obvious pathological change in the tissues from the affected plants. However, the xylem vessels in affected tissues seem to be of larger diameter as compared to those in healthy tissues (Shastry *et al.*, 1993). They through the SEM studies provided direct evidence of impaired water uptake in new wilt affected plants both under moisture stress and water logged conditions. They observed that emboli were formed in the xylem vessels of roots, stems and petioles of affected plants indicating that the xylem water comes under tension in response to reduced water uptake, the water column begins to rupture and forms gas filled emboli that render the vessels inactive. This seems to create an additional resistance to water transport to shoot. In revived plants, the vascular emboli are either reduced in size or totally disappear indicating that the recovery probably depends on transportation slow enough not to exceed the impaired ability of the vascular system to supply water to the shoot. Vascular repair presumably occurs because tension on the xylem water relieves, the gases in the vascular water redisolve in the nearby water and the water columns reconnect as the gases become fully dissolved.

Nutritional Imbalance

The nutrient status of leaves from wilt affected and healthy plants remains unchanged (Reddy *et al.*, 1984) however, it has been observed that immediately after the wilt the uptake of nutrients is reduced in the leaves (Anonymous, 1987; Krishna *et al.*, 1999; Pundarikakshudu, 1984, 1988). The elemental analysis of wilt affected plants by X–Ray fluorescence spectroscopy indicate that some inorganic elements particularly potassium and calcium were significantly less in roots of wilted than the healthy plants (Balsubramanya *et al.*, 1992). Shastry *et al.* (1993) have shown accumulation of N, P, K, Cu, Fe, Mn, Zn and Bo, while the calcium content was reduced in leaves. Similarly, an increase in total amino acids has been observed in seeds, while there was a decreasing trend in leaves. Methionine content tends to remain low both in seeds and leaves from wilted plants. Reversible leakiness and altered cell membrane permeability was found in wilt affected tissues of roots and leaves. It was proposed that the leakiness and alteration may be due to the transient reduction in the concentration of calcium which permits revival of some plants later when stress is relieved. New wilt was considered to be initiated by an elastic strain emanating from the presence of impaired water uptake/moisture stress.

Physico-chemical characters of the soil

Since the pathogenic nature of the wilt could not be established, several workers analysed the physico-chemical characteristics of soil and their involvement in the malady. Analysis of the soil samples from wilt affected plots revealed that there remained no appreciable difference in the composition with reference to macro-and micro-elements, while the pH of the soil ranged between 8.3 and 8.4 (Krishna *et al.*, 1988). Manganese injury or phosphate deficiency as the cause of wilt was ruled out by Reddy *et al.* (1984). Further analysis of soil and plant samples suggested that soil nature or nutrition was not the cause of wilt (Pundarikakshudu, 1984). Mudholkar (1984) suggested that certain specific conditions like high pH and sudden waterlogging of soil after a long dry spell could aggravate the incidence of wilt in cotton. Basu (1984) has observed that heavy soils and low soil temperature favour the disease. Shastry (1993) confirmed that physico–chemical characters of soil do not play any role in wilt development. Further, soil and foliar application of major (N, P and K) and minor elements (Cu, Mn, Fe, Zn and Bo) failed to control the disease. Soil application of Ca is also unable to revive the plants.

Ecoclimate

Environment seems to play an important role in the occurrence of the disease (Mandloi *et al.*, 1984; Raj *et al.*, 1984). Mandloi *et al.* (1984, 1985) and Krishna *et al.* (1988) noticed that the occurrence of wilt was more associated with weather shocks like dry spell, high temperature or continuous rains for a fortnight during August/September. They found that the occurrence was specifically associated with dry spells. Raj *et al.* (1990b, 1991) have, however, observed that prolonged drought was not a pre-requisite for disease onset, it plays an important role in aggravating the incidence. An ecoclimate conducive for the development of moisture stress (low rainfall and longer dry spell, higher maximum and minimum temperatures, low soil moisture and higher cumulative rate of change of maximum temperature) during the first 15 days of September is critical for the development of wilt (Shastry, 1993). Higher potential evapotranspiration values *i.e.* water limiting conditions suggested that PET values could be of use to forecast the occurrence of wilt. Two linear regression models have been developed to relate the occurrence of wilt with cumulative number of dry days and waterlogging duration in days as Y1=0.9362 + 0.1897 × 1 and Y=–1. 4013 + 0.7850 × 2 respectively, since the wilt develops under waterlogged conditions. Also, through SEM studies Shastry (1993) concluded that under both the situations (waterlogged and dry spell) there was impairment of water uptake and indicated that the principal morphological changes during New Wilt might be of adaptive value in terms of reducing further moisture stress effects under both the situations. These models can be used to have an idea of occurrence of wilt under prevailing weather conditions.

Role of ethylene

Possible involvement of excess ethylene production with the wilt has been demonstrated by Sahay *et al.* (1984) and Sahay (1988a). Sahay *et al.* (1988) and Shastry (1993) have also postulated that the effect of synthetic pyrethroids on the expression of wilt susceptible genotype is mediated through excess ethylene production. Artificial spray of ethylene on healthy plants induces wilt symptoms and the plants recover by the spray of silver nitrate due to neutralization of ethylene. Hence, it was concluded that the New Wilt was the result of high ethylene production in the plant (Sahay *et al.*, 1984). Depletion in methionine control both in seeds and leaves of wilt affected plants may be due to its conversion into ethylene as is known to occur in stressed plants where methionine acts as a precursor of ethylene. Krishna *et al.* (1988) have reported the presence of a number of fungi including *Penicillium* spp. In the rhizosphere of wilted plants which are known to produce ethylene in soils. They have suspected their role in excess ethylene production during the progress of wilt. However, occurrence of *Penicillium* spp. In soil is a common phenomenon and their presence alone cannot be attributed as a cause for the wilt (Balsubramanya *et al.*, 1992).

Synthetic pyrethroids and New Wilt

Application of synthetic pyrethroids induces quick wilt like symptoms in susceptible hybrids (Sahay, 1988b; Choulwar and Kharwade, 1989a; Mayee *et al.*, 1989; Chauhan *et al.*, 1989; Shastry, 1993). Shaw *et al.* (1989) have observed a linear relationship between the increase in the population of aphids and whiteflies and the incidence of wilt.

With the increase in the population of these insects, there was a corresponding increase in the incidence of New Wilt. It is interesting to note that the increase in the population was induced by the application of synthetic pyrethroids. Sahay (1988) and Shastry (1993) have postulated that the iatrogenic effect of synthetic pyrethroids may be mediated through excess ethylene production. Sahay (1988) considered synthetic pyrethroids to be a cause of wilt, while Shastry (1993) contradicted it because the malady was observed even in unsprayed crop. Mayee (1992) also found that even after removal of flowers, synthetic pyrethroids sprays increased parawilt incidence disproving the notion that creation of sink due to higher translocation of photosynthate to bolls as a consequence of effective boll retention is responsible for the wilt. In the absence of insecticide too, the wilt does occur indicating clearly that the pyrethroids cannot solely be implicated in the cause of New Wilt but can be an aggravating factor for the disease.

Inheritance

The incidence of New Wilt is very high particularly in hybrids JKHy–1, DCH–32 and MECH–1. The JKHy–1 and DCH–32 hybrids have one of the parents from acclimatized African *Gossypium hirsutum* race *punctatum* as direct parent or in the lineage of one. Some of the hybrids such as JKHy–1 and DCH–32 have Reba–B–50 and BJA–592, both from Africa, involved directly or indirectly as one of the parental lines. In the absence of pedigree for MECH-1. it was difficult to draw inference but could not be different from what is being contemplated (Bhale, 1984; Srinivasan, 1984). In spite of Reba–B–50 being used for more than a decade throughout the country, no report on the occurrence of wilt on Reba B 50 casts doubt on this hypothesis. This is also supported by the fact that the malady has been seen to occur on many other varieties which do not involve any of the African lines in their pedigree. Not only Reba–B–50 group but several other genotypes were also found to be affected. Progenies of crosses where these lines have been used also exhibit wilt symptoms. It was proposed that genetic factor alongwith the environment seemed to play an important role (Mandloi *et al.*, 1984, 1985; Raj *et al.*, 1984). Mandloi and Krishna (1988) have studied the pattern of inheritance of this wilt through a 10 × 10 diallel and 12 × 4 line x tester crosses and have proposed a modified duplicate factor model for its inheritance with following assumptions:

- The character is governed by two pairs of independent genes.
- Original population is double heterozygous for this character.
- The factor follows Mendelian laws of inheritance.
- There are equal chances of survival and random mating of gametes and survival of zygotes.
- Penetrance and expressivity of the character is governed by environment.

According to this proposition, the ratio of healthy and wilted plants in a population can be 16:0, 15:1, 14:2, 12:4, 8:8 and 0:16 but the common ratio will be 15:1 as healthy : wilted plants. This, therefore, explains the general occurrence of wilt in the fields around 6.25%. This proposition has been confirmed by Shastry (1993) by studying genetic parameters of variation and showed the role of additive gene action in expression of this wilt. A strong positive correlation exists between the wilt and number of bolls per plant. Higher number of bolls/plant coupled with environmental stress may predispose the plant to wilt. The wilt has been shown to be a genetically controlled physiological disorder expressed under stress (moisture stress or waterlogging) conditions of the environment.

Mathematical Modelling to Estimate Losses

Recently, attempts have been made to estimate the losses in yield due to this wilt. Krishna and his associates have evolved a mathematical model to estimate yield losses as:

$$Y_c = 1 - \frac{b_w N_w}{b_h N_h}$$

Where, Y_e = estimated loss in yielded, W = wilt percentage, N_w N_h and b_w b_h are the average number and weight of bolls on wilted and healthy plants, respectively, these estimates are based on the actual number and weight of bolls/plant in healthy and wilted plants and wilt incidence. An improved model has been developed by Shastry (1993) after considering the yield contribution of revived plants as: potential

$$\text{Potential yield}(Y_0) = \frac{Y_0}{100 - W_1 + W \times 0.58182} \times 100$$

Where, 0.58182 is a standardized constant, Y_0 and W_1 are observed yield and incidence of wilt (%), respectively. These models can be used to assess yield losses due to this disease.

Management

The efforts to devise suitable control measures have yielded very little success. Krishna *et al.* (1988) showed that 60–70% wilting plants recovered to normal by application of 2–3 solution of N+P of 0.5% per wilting plant if drenched as wilt initiation stage. Wilted plants have also shown to recover by application of Thimet+Carbendazim+ Metronidazole (Anonymous, 1990) and Metronidazole (Choulwar and Kharwade, 1989a; Mayee *et al.*, 1989; Anonymous, 1992). However, these results did not prove consistent enough to be applicable for field control.

References

Anonymous (1985a). Annual Report, AICCIP (1984–85). Regional Agricultural Research Station, Khandwa M.P.

Anonymous (1985b). Annual Report, AICCIP (1984–85). Marathwada Agricultural University, Parbhani, Maharashtra.

Anonymous (1986). Annual Report, AICCIP (1985–86). Regional Agricultural Research Station, Khandwa M.P.

Anonymous (1987). Annual Report, AICCIP (1986–87). Regional Agricultural Research Station, Khandwa M.P.

Anonymous (1989). Biennial Progress Report, AICCIP (1987–89). Agricultural Research Station, Sriganganagar, Rajasthan.

Anonymous (1990). Annual Report, AICCIP (1989–90). Marathwada Agricultural University, Parbhani, M.S.

Anonymous (1992). Annual Report, AICCIP (1991–92). Mahatama Phule Krishi Vidyapeeth, Rahuri, M.S.

Anonymous (2001). All India Coordinated cotton improvement project. Annual report, Khandwa. p. 85.

Anonymous (2004). All India coordinated cotton improvement project. Annual Report, Coimbatore. pp. 2–6.

Ahuja, S.L. (2003). Inter-relationship and variability analysis on area,production and yield in major cotton producing countries of the world. *J. Cotton Res. Dev.* 17(1): 75–85.

Balsubramanya, R.H., Paralikar, K.M., Chidambareswaran, P.K., Patil, N.B. and Sundaram, V. (1992). A note on the New Wilt of cotton. *J. Indian Soc. Cott. Improve.*, 17(1): 74.

Basu, A.K. (1984). Unknown wilt of cotton. Proc. Group Discussion on New Wilt in Cotton, Nov. 6–7. CICR, Nagpur. p. 44.

Basu, A.K., Narayanan, S.S., Deshmukh, R.K. and Mukewar, P.M. (1992). Linkage of crop improvement programmes with seed technological research in cotton. *Seed Tech. News.*, 22: 10–12.

Bhale, N.L. (1984). On new disease of cotton. Proc. Group Discussion on New Wilt in Cotton Nov. 6–7. CICR. p. 9.

Chauhan, M.S., Sharma, P.D. and Kairon, M.S. (1989). Field observations on sudden wilting and drying of cotton plants-an unsolved problem. Paper presented at national Seminar on Futuristic Approaches in Cotton Improvement, Hisar. Feb. 24–25. p. 56.

Choulwar, A.B. and Kharwade, L.R. (1989). Iatrogenic influences of synthetic pyrethroids on parawilt of cotton. *Indian Phytopath.*, 42: 352.

Deshmukh, L.D., Ansingkar, A.S. and Thombre, P.G. (1989). Economical significance of parawilt in relation with fibre qualities. National Seminar on Futuristic Approaches in Cotton Improvement, Hissar. Feb. 24–25. p. 6.

Ebbels, D.L. (1975). Fussarium wilt of cotton. A review with special reference to Tanzania. *Cotton Growers Rev.,* 52(4): 295.

Ekbote, M.V. (1984). A short note on cotton transitory wilt. Proc. Group Discussion on New Wilt in Cotton, Nov. 6–7. CICR, Nagpur. p. 22.

Jadhav, S.B., Banerjee, S. and Julka, R. (1992). An assessment of fibre quality of cotton affected by New Wilt phenomenon. *J. Indian Soc. Cott. Improve.,* 17(2): 133.

Jadhav, R.Y., Kadam, S.B., Mehetre, S.S. and Patil, S.D. (2002). Tillage and nutrients for summer-irrigated cotton proceeded by sugarcane in Western *Maharashtra. J. Cotton Res. Dev.,* 16(2): 190–192.

Jayaswal, A.P. (1984). A note on wilting in cotton possibilities of involvement of insects. Proc. Group Discussion in New Wilt of Cotton, Nov. 6–7. CICR Nagpur. p. 32.

Kairon, M.S., Rao, M.R.K., Deshmukh, R.K. and Mukewar, P.M. (1997). Souvenir-National group meeting on cotton seed. pp. 48-51. *Central institute for cotton research, Nagapur*, January, 21–22. 1997.

Krishna, A., Mandloi, K.C. and Tiwari, R.J. (1988). New Wilt of cotton: A physiological disorder. All India Cotton Production and Utilization Conference, April 23–24. CTRL, Bombay. p. 36.

Mali, V.R., Mayee, C.D., Kore, S.S. and Jadhav, V.T. (1984). A parawilt of hybrid cotton in Marathwada region of Maharashtra. Proc. Group Discussion on New Wilt in Cotton. Nov. 6–7 CICR, Nagpur. p. 56.

Mandloi, K.C. (1979). Field investigation report, 1979. Mandloi, K.C., Julka, R., Krishna, A., Shroff, V.N. and Kotwani, H.G. 1984. Wilters in cotton. Proc. Group Discussion on New Wilt in Cotton, Nov. 6–7. CICR, Nagpur. p. 47.

Mandloi, K.C., Julka, R., Krishna, A., Shroff, V.N. and Kotwani, H.G. (1985). An unidentified wilt like disease on cotton. *J. Indian Soc. Cott. Improve.*, 10(2): 117.

Mandloi, K.C. and Krishna, A. (1988). Duplicate factor model for New Wilt of cotton. All India Cotton Production and Utilization Conference. April 23–24, CTRL, Bombay. p. 19.

Mandloi, K.C., Nagwekar, S.N., Pai, S.D., Oka, P.G., Krishna, A. and Kotwani, H.G. (1987). The New Wilt and its effect on fibre quality in cotton. *J. Indian Soc. Cott. Improve.,* 12(1): 44.

Sahay, R.K., Bhale, N.L., Pundarikakshudu, R. and Gawande, M.C. (1984). Physiological studies on New cotton wilt. Proc. Group Discussion on New Wilt in Cotton, Nov. 6–7. CICR, Nagpur. p. 33.

Shastry, P.P. (1993). Studies on the New Wilt of cotton. Ph.D. thesis submitted to the JNKVV, Jabalpur.

Sharma, R.A., Thakur, H.S., Holkar, S., Chauhan, S.S. and Devkant (2003). Impact of tillage, land treatments and organic residue management on soil health, drainage and crop productivity of rainfed cotton", Res. Bull. No. AICRPDA-NATP-RCPS-11/2003. Published by AICRPDA, Indore, M.P. p. 32.

Shaw, S.S., Das S.B., Verma, R.S. and Mandloi, K.C. (1989). A note on the resurgence of homopterans and wilting in rainfed cotton. *J. Cott. Res. and Dev.* 3(1): 83.

Sheo, Raj. and Taneja, N.K. (1987). Diseases of cotton and their management. *In*: Plant protection in field crops. *Eds*. Rao, M.V. and Sithanathum, S. Plant Protection Association of India, Hyderabad, India. pp. 301–17.

Shreenivasan, S. (2004). On the competitiveness of Indian cottons on quality front in the free market area. Paper presented in international symposium on "Strategies for sustainable cotton production. A Global vision", Proc. 1 crop improvement pp. 5–8. Nov. 2004, held at UAS, Dharwad.

Srinivasan, K.V. (1981). Cotton diseases in India. Present knowledge and need for further research. Biennial Workshop, AICCIP (1979–81), MAU, Parbhani.

Srinivasan, K.V. (1984). The New Wilt disease affecting cotton. Proct. Group Discussion on New Wilt in Cotton. Nov. 6–7. CICR, Nagpur. p. 9.

Swaminathan, M.S. (1999). Cotton and Rural Livelihoods. Silver Jubilee Lecture Series. Indian Society for Cotton Improvement, Mumbai, p. 28.

14

GM Crops

Traditional methods for improving traits in a crop plant has been practiced for ages and no doubt has delivered but is very time consuming. Genetic engineering allow for specific changes to be made quickly, on a molecular level. The world population is expected to grow rapidly, particularly in developing countries the amount of available agricultural land is limited. Only 10 percent of the world's land surface is arable, and intensive farming and soil erosion are growing problems in some areas. It will be difficult to feed this growing population if revolutionary approach to increase the production and productivity is not adopted. To overcome these problems revolutionary approach is needed to increase the overall production of food grains in limited arable land. Biotechnology offers a partial solution- "GM technology, coupled with important developments in other areas, should be used to increase the production of main food staples, improve the efficiency of production, reduce the environmental impact of agriculture, and provide access to food for small–scale farmers." Biotechnology already is beginning to make a contribution. Maize, sorghum and wheat are being developed to provide more lysine, an important dietary protein. "Golden rice," enriched with beta carotene, will help combat vitamin-A deficiency. Further, scientists have enriched the same strain of rice with additional iron to combat anemia. Similarly "golden mustard" is also being developed to yield oil with high levels of provitamin A. Foods with fruits and vegetables with improved shelf life are in the offing. Crops resistant to drought and saline soils are expected to reach the farmers very soon. This developments are indicators of biotechnology playing a big role in mitigating the problems posed by increasing population and decreasing arable land (Clive James, 2010).

Genetically Modified Crops were introduced for commercial farming in 1996 and 2011 marks the sixteenth anniversary of the commercialization of biotech crops. The experience of almost one and half decade has been very promising and as promised has helped in improving the yields and a big step forward in reducing agro-chemicals, thus reducing the cost of cultivation on one hand and simultaneously delivering socio–economic and health benefits. The rate of adoption of biotech crops has been spectacular and no other technology can match this rate of growth. Between 1996 and 2009, developing and industrial countries contributed to a record 80-fold increase in the global area of biotech crop from 1.7 million hectares in 1996 to 134 million hectares in 2009. Public sector institutions have also contributed significantly in this growth. The problems of weeds, insect pests and diseases have been addressed successfully (Clive James, 2010).

The other side of the coin highlights the continuing debate on biotech crops particularly in country of the European Union (EU), the opposition of *Bt* Brinjal in India but despite this, millions of large and small farmer in both industrial and developing countries have continued to increase their planting of biotech crops by double–digit adoption growth rates almost every year since 1996, indicating the tremendous benefits harnessed by the biotech farmers. Nearly 14 million farmers in 25 countries grew biotech crops in 2009 (James, 2009b).

Genetically modified crops can significantly contribute in the poverty alleviation program [Millennium Development Goals (MDG)] aimed to lower poverty, hunger and malnutrition by half by 2015. GM crops can help in increasing crop productivity, assisting nations to become self–sufficient, Conserving biodiversity, reducing the environmental footprint of agriculture, mitigating some of the challenges associated with climate (increased frequency and severity of droughts, floods, changes in temperature, rising sea levels exacerbating salinity and changes in temperature) and reducing greenhouse gases, increasing stability of productivity and production, the improvement of economic, health and social benefits and thus provide significant and important multiple and mutual benefits to producers and global society.

In 2010, the global area of biotech crops increased at a rate of 10% reaching covering over 148 million hectares (Table 1). Again, the number of farmers growing biotech crops rose to 15.4 million. An independent study published by the US National Research Council (http://www.nep.edu) is entitled "The impact of genetically (GE) crops on farm sustainability in the United states. "The study concluded that "many US farmers are realizing substantial economic and environmental benefits such as lower production cost benefits, fewer pest problems, reduced use of pesticides and better yields compared with conventional crops." The study however, noted that biotech crops tolerant to `glyphosate could develop more weed problems as weeds evolve their own resistance to glyphosate and that herbicide crops could loose their effectiveness unless farmers also use other proven weed and insect management practices. Similarly, Brazil is estimated to have enhanced farm income from biotech crops by US$3.5 billion between 2003 and 2009 (Clive James, 2010).

Table 1: Area under Genetically Modified Crops in 2010

Year	*Hectares (million)*	*Acres (million)*
1996	1.7	4.3
1997	11.0	27.5
1998	27.8	69.5
1999	39.9	98.6
2000	44.2	109.2
2001	52.6	130.0
2002	58.7	145.0
2003	67.7	167.2
2004	81.0	200.0
2005	90.0	222.0
2006	102.0	252.0
2007	114.3	282.0
2008	125.0	308.8
2009	134.0	335.0
2010	148.0	365.0
Total	1,097.9	2,716.0

Source: Clive James, 2010

A detailed analysis (Clive James, 2010), estimated that the total global benefits from RR soybean in Argentina for the first 10 years of commercialization, 1996 to 2005 was US$46 billion . Argentina gained approximately US$20 billion indirect benefits from RR soybean between 1996 and 2005. The study indicated that the benefits were a direct result of the adoption of RR soyabean. The adoption rate was very high with the introduction of herbicide tolerant soyabean in Argentina in 1996 and now covering 99% of the total area. The production cost is also lowered. In India with the introduction of *Bt* cotton, both the production and productivity of cotton between 2002 and 2010, the average yield of cotton, which used to be one of the lowest in the world, increased from 308 kg per hectare in the 2001–02, to 526 kg per hectare in 2008–09 with 50% or more of the increase in yield, attributed to *Bt* cotton. Thus *Bt* cotton is a major factor contributing to higher cotton production which increased from 15.8 million bales in 2001–02 to 24.4 million bales in 2005–2006, 28 million bales in 2006–2007 and 31.5 million bales in 2007–08, which was a record cotton crop for India (Cotton Advisory Board, 2008). In India the private seed sector is focusing only on cash crops like cotton and high– value, low volume segments such as vegetables and horticulture crops. The private seed industry's role in promoting genetically modified (*Bt*) cotton has been particularly significant. India is now mega cotton producing country in the world as noted in the Economic Survey of 2006–07.

The phenomenal increase in cotton production during the last decade, India has transformed itself from a net importer to a net exporter of cotton. Export of cotton has registered a sharp increase from a paltry 0.05 million bales in 2001–02 to 5.8 million bales in 2006–07 before touching a high of 8.8 million bales in 2007–08 (PIB, 2007). In 2009–10, cotton export was to a tune of 8.3 million bales.

The number of events, as well as the number of *Bt* cotton hybrid and companies marketing approved hybrids in India has been on an increase since its introduction in 2002. In 2010, the number of *Bt* cotton hybrids that were permitted for commercial cultivation was 780. This increase in number of hybrids has provided much more choice year after year to farmers in the North, Central and Southern regions, where specific hybrids have been approved for cultivation in specific regions (Tables 2 & 3). As far as the events are concerned, a total of six events were approved for incorporation in a total of 780 hybrids with one event from the public sector (BN *Bt* in Bikaneri Nerma (BN) and hybrid NHH–44). The sixth event MLS–9124 was approved for the first time in 2009 (Table 4).

There are some new events (Tables 4, 5 & 6) in cotton that are in the process of testing in the field. In 2010, four new cotton events expressing insect resistance and herbicide tolerance in the cotton plant have been tested. Normally farmers have been using physical

Table 2: Status of *Bt* cotton in India, 2010

Deployment of approved *Bt* cotton events/ hybrids/variety by companies/institutions in India, 2002 to 2010									
Event	***2002***	***2003***	***2004***	***2005***	***2006***	***2007***	***2008***	***2009***	***2010***
NORTH ZONE Haryana Punjab Rajasthan				6 Hybrids 1 Event 3 Companies	14 Hybrids 3 Event 6 Companies	32 Hybrids 4 Event 14 Companies	62 Hybrids 4 Event 15 Companies	164 Hybrids 5 Event 26 Companies	271 Hybrids 5 Event 31 Companies
CENTRAL ZONE Gujarat Madhya Pradesh Maharashtra	3 Hybrids	3 Hybrids	4 Hybrids	12 Hybrids 1 Event 4 Companies	36 Hybrids 4 Event 15 Companies	84 Hybrids 4 Event 23 Companies	148 Hybrids 4 Event 27 Companies	296 Hybrids 6 Event 35 Companies	459 Hybrids 6 Event 35 Companies
SOUTH ZONE Andhra Pradesh Karnataka Tamil Nadu	3 Hybrids	3 Hybrids	4 Hybrids	9 Hybrids 1 Event 3 Companies	31 Hybrids 4 Event 13 Companies	70 Hybrids 4 Event 12 Companies	149 Hybrids 4 Event 27 Companies	294 Hybrids 6 Event 35 Companies	444 Hybrids 6 Event 35 Companies
Summary									
Total of hybrids	3	3	4	20	62	131	274	522*	780*
Total of events	1	1	1	1	4	4	4	6	6
Total companies	1	1	1	3	15	24	30	35	35

Source: Clive James, 2010.

and cultural methods for weed control in cotton fields. In Madhya Pradesh, Maharashtra and Gujarat a mechanical implement (Kolpa/ Dora) is being effectively used both for weed control and interculture. However, during periods of incessant rains it becomes practically impossible to control weeds. Under these circumstances hybrids stacked with gene for herbicide tolerance becomes very useful. Further availability of labour for manual weeding is not proving to be cost effective due to increasing wages and non–vaailability of sufficient labour. In North India there is increasing trend in the use of herbicides for weed control in cotton. However; extreme care is required particularly while using non-selective herbicides. These new events are undergoing field testing and pending approval have the potential to be commercialized sometime between 2011 and 2015 (Tables 2 & 3) (Clive James, 2010).

Table 3: Important landmarks in cotton improvement India, 1950 to 2010

Year	*Technology introductions*
1950s–60s	Improvement of desi cotton.
1960s–70s	Introduction of *Gossypium* varieties of cotton
1970s–80s	Development of cotton hybrid development from intra –species and interspecies combinations.
1980s–90s	Introduction of pyrethroid-based insect control mechanisms
2002–2010	Commercialization of *Bt* technology (with single as well as combination of genes) in cotton hybrids
2010–2015	A milestone period marking the introduction of stacked *Bt* / HT technology in cotton improvement of fiber quality and development of marker aided selection (MAS) based high - yielding cotton genotype .A major innovation in breeding and cotton agronomy by introducing high – density cotton genotypes and agronomic practices to boost cotton yield per unit area State –of –the –art techniques/mechanization for cotton picking ginnig and processing
2015 Onward	Deployment of triple gene *Bt* technology, sucking pest resistant and drought tolerant cotton. Stacking of insect resistant, sucking pest resistant and drought tolerant cotton traits

Source: Clive James, 2010

There is wide variation in the use of various events both in geographical distribution and extent of use. The first event MON531 sourced from Monsanto by MAHYCO Hybrid Seed Company was commercialized in 2002 appears to be gradually being phased out.

The seventh event, developed by Mahyco and sourced from Monsanto, featuring stacking of two events (including insect resistance and herbicide tolerance in cotton) expressing three genes cry1Ac and cry2Ab to confer insect resistance and EPSPS genes to impart herbicide tolerance are being field tested. In 2010 four cotton hybrids including two hybrids for North zone and two for Central and South zones were approved for seed production in an area of 25 acres per hybrid and is waiting for approval for commercial in 2011–12. Similarly, eighth event has been developed by Dow Agro Sciences, expressing multiple genes including *cry1F*gene and (Event 3006–2010–23 and event 281–24–236) and has two genes for insect protection. The ninth event , known as a combination of -1x -24 is being developed by JK Seeds featuring two genes *cry1Ac* and *cry 1Ec*, sourced from IIT Kharagpur and NBRI Lucknow, India Hybrid expressing Event–1 x Event–24 have been field tested in the North, Central and South zones in 2010. The tenth event named, glytol developed by Bayer Biosciences and contains EPSPS gene conferring tolerance to cotton hybrid sprayed with the herbicide Glyphosate is being field tested for further selection.

Genetic engineering is also aiming to develop crop plants that contain nutrients known to help fight disease or malnourishment. The classical example in this direction is the development of *Golden Rice*, which contains beta-carotene, the precursor for Vitamin A. Three genes, two from daffodils and one from a bacterium have been integrated into rice plants to achieve this. Scientists have also realized that abiotic stress resistance will help in increasing the area under arable land particularly in developing countries. In this, direction crop plants with tolerance to conditions like salinity, cold and drought. Sufficient progress has been made to achieve this and it is expected to be available for commercial use during the next few years. It is interesting to note that a Canadian company Nexia announced development of transgenic goats that produced spider silk proteins in their milk which is the strongest fiber known to man, stronger than kevlar (used to make bullet-proof vests), with a higher tensile strength than steel.

Application Potential of Biotechnology in Agriculture

Genetic engineering provides an opportunity to modify genetic composition of plants, animals, and microorganisms beyond our imagination. In this context judicious usage of this technology will go a long way in mitigating hunger and improving health of the growing population.

Herbicide tolerance allows crops to withstand injury from herbicides. Non-selective herbicides kill virtually all plants and cannot be used on crops. Incorporating genes for tolerance to select herbicides in crops will help in economic and effective weed control. The current set of commercially available herbicide-tolerant crops is tolerant to three herbicides based on three active ingredients, bromoxynil, glyphosate, and glufosinate. Similarly, all of the commercially available insect–tolerant plants contain a version of the toxin *Bacillus thuringiensis* (*Bt*), which is found in nature in soil bacteria. *Bt* toxins are highly effective for many pest organisms, like beetles and moth larva, but not toxic to mammals and most other nontarget organisms. A major concern among farmers and environmentalists is that wide use of *Bt* crops will lead to the rapid development of resistance. Planting of refugia should be either made compulsory (implementation of law??) or alternative refugia systems need to be developed which are acceptable to the farming community. Cultivation of transgenic cotton is not permissible for organic cultivation; there is a problem of contamination. In view of this a debate should be initiated on permitting transgenic cotton which actually reduces the usage of chemicals. Another application of biotechnology to crops is virus tolerance. A novel approach is to incorporate a gene from the virus itself that is expected to produce certain viral proteins which helps in defending the crop plants against infections.

Many other genetically engineered products have been envisioned, but only a few have so far come to market. Several tomatoes engineered to delay ripening have been approved for commercial use. In some cases, delayed ripening just prolongs shelf life. But for the FlavrSavrTM, the objective was to increase the time on the vine without softening, producing a transportable and tasty winter tomato. After a highly publicized rollout, the FlavrSavrTM too is off the market. The problem appears to have been with transportability rather than taste.

On the livestock side, a drug has been produced for dairy cows–recombinant Bovine Growth Hormone (BGH) or Bovine Somatotropin (BST)–by engineering a bacterium to contain the gene for the hormone. The drug is administered to cows to increase milk production, despite the chronic oversupply of milk in the United States. A highly controversial product, BGH is currently used on about a third of the U.S. dairy herd.

An interesting product not related to agriculture is a rabies vaccine intended for use on wild raccoons. In this case, genetic engineering was used to construct a "hybrid" virus made up of a component of the rabies virus inserted into an unrelated "carrier" virus. The resulting virus confers immunity to rabies but poses no danger of causing the disease. Baits laced with the vaccine have been distributed in many parts of the eastern United States in attempts to combat rabies in wild raccoon populations. The vaccine has been approved by the U.S. Department of Agriculture, despite suspicions that it has been only marginally, if at all, effective. Early studies on efficacy failed to demonstrate that the product could control rabies in wild raccoon populations. Data from more recent studies are being withheld from the public as confidential business information.

There is a lot of skepticism about a technology known as Terminator Technology. Terminator technology refers to research of seeds/plants that produce sterile seeds. At present, this technology has not been used and has not been exploited for commercial use. In the future, this technology could be used to prevent any gene flow between biotechnology and traditional crops.

Terminator Technology for Transgenic Crops

The most contentious issue in respect of the application of genetic engineering particularly in agriculture if "terminator technology" commonly refered to as 'Technology Protection System' by those who intend to commercially exploit this technology. On the 3rd of the March 1998 The US Department of Agriculture and cotton seed breeders Delta and Pine Land Company acquired US patent 5,723,765 for their "Technology Protection System" (TPS). On September 15th 1998 Astra Zeneca received US patent 5,808,034 on it's own terminator type technology. This was branded the "verminator" as the genes were taken from brown rats. The approach is aimed is to kill only the embryos–leaving other important seed components such as oils and proteins intact.

There are several ways to exploit the technology. It essentially involves three steps to express the phenomenon by which the plants will produce apparently healthy seeds but will be sterile or will not germinate and thus protect the basic technology. This technology offers no benefit for farmers or consumers. The only advantage would be to commercial seed companies hoping to force farmers back to market each season to purchase seeds. The potential threat to small farmers has caused widespread outrage and protests. Indian farmers who uprooted and burned Monsanto's GE cotton test plots were fuelled by their anger about the terminator, even though these were not themselves terminator crops. There are 100 million farmers in India and 80% are dependent on farm-saved seed. In December 1998 the Indian Agricultural Minister Som Pal was forced to assure the Upper House that terminator would not be allowed in India. The technology is expected to be commercially exploited by adding terminator genes to the crop. An inducer is added to the seed before it is sown. As a result of this the farmers will harvest mature but sterile seeds. The engineered gene comes into play very late in seed development when a special switch under the control of the inducer turns on the gene causing it to produce toxin. The toxin kills the embryo that is part of each mature seed.

Following opposition from many quarters most of the Multinational seed companies have expressed their willingness not to use this technology. This willingness to abandon the terminator may have something to do with the possibility of developing engineered plants whose traits are controlled by a "switchable promoter". This system allows traits to be switched on or off by application of a chemical catalyst. Sterility is just one of the traits which could be under the control of such promoters. Astra–Zeneca has at least seven patent applications on different promoter systems. This technology could be used to create plants whose desirable traits would be switched on only by the application of a chemical. Farmers will have to buy seed and the chemical that goes with it. What could be more logical for companies than to put the required chemical in its own herbicide or pesticide? GE companies will be able to produce seed that contains multiple GE traits; farmers will have to pay for the chemical to activate each specific trait. RAFI who originally coined the name terminator call this "traitor technology".

References

Clive, James (2009). Global Status of commercialized Biotech/GM crops 2009. ISAAA Brief No. 41. ISAAA, Ithaca, NY.

Clive, James (2010). Global Status of commercialized Biotech/GM crops 2010. ISAAA Brief No. 42. ISAAA: Ithaca, NY.

Shastry, P.P., Mishra, P.K., Sharma, R.A. and Tomar, V.S. (2006). Cotton Production Technology for Madhya Pradesh. Tech. Bulletin No. DRS/JNKVV/2006/01.

http://www.goldenrice.org/PDFs/ The Golden Rice Tale.pdf.

http://www.nep.edu

Times of London, November 4, 1998; and Times of India, December 4, 1998.

Oliver, M.J. (1998). "Control of plant gene expression," US Patent Number 5,723,765.

Crouch, M.L. (1998). "How the Terminator terminates an explanation for the non-scientist of a remarkable patent for killing second generation seeds of crop plants." The Edmonds Institute, Edmonds, Wash.

(1998). Union of Concerned Scientists. "Biobit and Terminator Technology." The Gene Exchange, Winter. p. 4.

www.rafi.org- For information on Terminator Technology.

Golden rice

Subject Index